Macroeconomics, Agriculture, and Exchange Rates

Macroeconomics, Agriculture, and Exchange Rates

EDITED BY

Philip L. Paarlberg
and Robert G. Chambers

Routledge
Taylor & Francis Group
LONDON AND NEW YORK

First published 1988 by Westview Press, Inc.

Published 2018 by Routledge
52 Vanderbilt Avenue, New York, NY 10017
2 Park Square, Milton Park, Abingdon, Oxon OX14 4RN

Routledge is an imprint of the Taylor & Francis Group, an informa business

Library of Congress Cataloging-in-Publication Data
Macroeconomics, agriculture, and exchange rates / edited by Philip L. Paarlberg and Robert Chambers.
p. cm.—(Westview special studies in international economics and business)
ISBN 0-8133-7562-2
1. Agriculture—Economic aspects—United States. 2. Foreign exchange. 3. Macroeconomics. I. Paarlberg, P. L. II. Chambers, Robert G. III. Series.
HD 1761.M33 1988
338.1'0973—dc19 **88-5653**
CIP

ISBN 13: 978-0-367-01467-4 (hbk)

CONTENTS

PART III
LINKAGES IN DEVELOPING COUNTRIES

PREFACE

The papers in this book are largely the result of a conference on the linkages between macroeconomics and agricultural trade sponsored by the International Agricultural Trade Research Consortium during the summer of 1986. Although the papers cover a wide variety of topics, they are interrelated. The first paper, by Chambers, surveys the current state of research by agricultural economists on the linkages between agriculture and the macroeconomy, especially the linkages through the monetary sector. He provides a critical history of past research efforts to understand the interaction between the money supply, exchange rate, interest rate, and the agricultural sector. This paper provides a foundation for those which follow.

The paper by Frankel and Froot, and that by McKinnon clarify that exchange rate determination is more than an equilibrium between demand for and supply of dollars arising from the trade account. These papers present a more modern view of exchange rate determination which developed in the 1970s following the collapse of the Bretton Woods Agreement. Frankel and Froot review modern theories of exchange rate determination. These theories explain the appreciation of the U.S. dollar in the early 1980s, but do not explain the persistence of that rise. They then expand the theory to explain the persistence of the high dollar due to the conflicting expectations of fundamentalists and technical analysts. McKinnon expands the treatment of the exchange rate by developing a linkage between the aggregate price level, the exchange rate, and global stock of money. He argues that spillover effects in the money market call for a revision of decision rules by monetary authorities.

The first three chapters establish some of the fundamental influences on the exchange rate. The next set of papers develop linkages between the macroeconomy and agriculture using traditional models. Shei and Thompson clarify the linkages through which inflation is transmitted among real sectors, including agriculture, and the monetary sector. Their analysis suggests that the increase in the money supply during the early 1970s had large impacts upon the rise in agricultural prices during 1973. They compare the monetary shock to shocks specific to the agricultural sector and argue that the monetary shock was a major contributor to the agricultural price rise. Shei and Thompson argue that agricultural prices overshoot their long-run equilibrium due to rigidities in other prices and output. The paper by Stamoulis and Rausser formally considers the issue of overshooting by treating agricultural markets as flex-price markets and the other sectors of the economy as fix-price markets. They demonstrate the necessary conditions for over- and undershooting. The empirical results support the analysis of Shei and Thompson concerning the importance of monetary policy for agricultural prices but do not confirm the hypothesis that agricultural prices overshoot their long-run equilibrium. The third paper in this section, by Just, considers some adverse consequences for agriculture associated with flexible exchange rates and macroeconomic instability. These adverse effects include overshooting, capital market imperfections and risk aversion. Just argues that the effects should be considered in macroeconomic policy formation. Deficiencies in current research prevent that information from being available to policy makers. The section ends with remarks by Robert L. Thompson, Assistant Secretary of Agriculture for Economics, on the interactions between U.S. agriculture and the macroeconomy as seen by a policy maker. Thompson briefly comments on the policy importance of many of the issues raised by previous papers. He also introduces two issues which received less attention because of the international focus of the conference--the farm financial crisis and U.S. fiscal policy.

The brief remarks by Schuh set the stage for considering macroeconomic-agriculture linkages in developing countries. Schuh considers the costs and benefits of the new macroeconomic agenda facing agriculture, but with emphasis on developing country effects. Schuh is concerned about issues such as instability effects, adverse changes in income distribution, and problems of agricultural adjustments to macroeconomic forces. These issues are often ignored if macroeconomic linkages to agriculture are analyzed from a developed country bias.

A specific case study of the problems associated with the new economic environment facing world agriculture, particularly the developing nations, is presented by Shane and Stallings. They review the consequences of the international debt crisis. Many of the externalities cited by Just and the problems noted by Schuh appear in the analysis of countries struggling to repay international obligations in the face of world recession, high interest rates, an appreciating U.S. dollar, and declining petro-dollar funds.

The early papers which modeled the linkages between agriculture and the macroeconomy focused on developed countries, especially the United States. Abbott argues that the macroeconomic models for developed countries are inappropriate for the developing world. This is a result of institutional structures which alter the mechanisms through which agriculture and the macroeconomy interact. Unlike developed countries, in many instances the agricultural sector of developing countries is the macroeconomy. In addition, the lack of monetary and financial markets in developing countries preclude many of the adjustment alternatives available in more sophisticated economies. Thus, when modeling the macroeconomic-agricultural linkages in developing countries, these structural constraints must be considered.

In addition to the efforts of the authors of the papers, three other people were critical to the success of the conference and this book. Professor Alex McCalla of the University of California--Davis made the arrangements for the conference facilities at Tahoe as well as providing a liaison between the Executive Committee of the International Agricultural Trade Research Consortium and the conference co-chairs. Laura Bipes of the University of Minnesota handled the administrative details of the conference and book, and further supervised the word processing. Linda Schwartz also at the University of Minnesota was responsible for the word processing and the high quality of the output is a direct result of this effort. Without the contributions of these three individuals this book would not have been possible.

Philip L. Paarlberg
Purdue University

Robert G. Chambers
University of Maryland

1

An Overview of Exchange Rates and Macroeconomic Effects on Agriculture

Robert G. Chambers

My task is to provide a backdrop for the papers that follow on the effects of exchange rates and related macroeconomic phenomena on U.S. agriculture. What I offer is a critical survey of some previous modeling attempts by a number of agricultural economists. The hypothesis that exchange rates and macroeconomic phenomena have important implications for U.S. agriculture has come full circle from a hotly disputed and, in many quarters, a rejected hypothesis to one widely accepted by academic and nonacademic agricultural economists. People continue to argue about magnitudes, of course, but there appears to be common agreement that it is important. The change that occurred, however, was not due largely to the force of the logical arguments offered by either side to the debate. Rather, perceived reality converted many confirmed disbelievers to the "enlightened" path. For this reason, it is worthwhile to review the last decade or so in agricultural trade to try to discover the reasons for this change.

AGRICULTURAL TRADE IN THE 1970s AND 1980s

A commonly accepted hypothesis is that U.S. agriculture's exposure to world markets has changed dramatically over the last 15 years. The value of U.S. agricultural exports rose from about $9 billion in 1972 to over $43 billion in 1981. The share of gross farm receipts coming from exports went from less than 15 percent to

---

Robert G. Chambers, Professor, Department of Agricultural and Resource Economics, University of Maryland, College Park, MD.

nearly 30 percent, and by 1981, some 36 percent of agricultural production was being exported. The percentages for the major export crops (wheat, corn, and soybeans) were even higher.

Since 1981, however, the bottom has literally fallen out of the U.S. agricultural export market. Having peaked at $43 billion in 1981, agricultural exports for 1986 were down to around $26 billion. Not only value, but volume as well has fallen. At the same time, our share of the world market fell for several important export commodities.

What caused the tremendous surge and latter slump in U.S. agricultural exports? No definitive answer is yet, or likely to be, forthcoming. However, a summary of conventional wisdom might run something like this: In the 1970s, the United States twice devalued and ultimately floated the dollar. Consequently, the dollar was a relatively weak currency during the 1970s which naturally made U.S. exports more attractive. Almost simultaneously, the Soviet Union initiated large grain purchases in world markets, while rising affluence and increased borrowing by developing countries increased demand for food commodities (LDCs became our fastest growing export market). The world agricultural trade economy seemingly expanded at a rapid rate, and the United States was well poised to take advantage of this expansion both because of its relatively cheap currency and because it could rapidly expand production.

By the late 1970s, when the rest of the U.S. economy was in the doldrums, some agricultural experts were touting the agricultural experience as an example for the rest of the economy. There was even an organized effort, taken quite seriously at the time, to use agricultural exports to lead the rest of the economy onto the golden turnpike of perpetual and steady growth. To many, the market for agricultural exports was unlimited and to some, even ever expanding.

Things turned sour in the 1980s. Agricultural exports plummeted. The fact that developments in the 1980s were the mirror image of what happened in the 1970s convinced many that exchange rates and macroeconomic policies did matter for agriculture. People who had been devoted skeptics or agnostics suddenly couldn't agree fast enough that the problem with agriculture had its foundations in the Federal Reserve's (Paul Volcker's) management of the money supply. The decision of the Federal Reserve to move from targeting interest rates to targeting money growth rates was seen as the death knell of American agricultural exports and perhaps even of American agriculture. However, like all monetary phenomena, it operated with a lag and the hue and cry didn't really start to rise until 1983 when much of the damage had been done. Among agricultural economists, however, it became an article of faith that was almost as soundly grounded as the belief in inelastic demand. The belief reached such an emotional level that one prominent researcher was heard to denounce Paul Volcker at a policy conference (I paraphrase here) as a murderer of baby pigs and cattle. If this sounds populist, it is, and the level of the debate in many agricultural circles started to assume a tenor that was curiously familiar to students of American history. The famous William Jennings Bryan "Cross of Gold" speech comes immediately to mind.

With all the rhetoric and unfounded assertions scattered about, the real policy problem of what to do about agricultural exports remains. And solving this problem requires an understanding of just what has caused the slump in exports. The papers contained in this volume are an attempt to examine just one cause, but what many agree is a candidate for the most important aspect of that problem.

In what follows, I intend to review first the literature on the effects of exchange rates upon agriculture and then turn to an overview of the literature of the effects of macroeconomic policy on agriculture.

EXCHANGE RATES AND AGRICULTURE

Without a doubt, the most important paper on exchange rates, other macroeconomic phenomena, and agriculture is the classic paper by Ed Schuh entitled "The Exchange Rate and U.S. Agriculture." What makes this paper so important is not the sophistication of the analytical argument. Rather, like all great papers, this one perceived a problem and the implications of that problem well in advance of the rest of the profession. What Ed Schuh pointed out in 1974 was immediately controversial and was even immediately disparaged by some, but as this paper is being written, it has become an important part of the conventional wisdom in agricultural economics. Perhaps, the best summary of this paper is the highest compliment one can pay to an academic paper: it opened a whole new field of inquiry. Dozens of Ph.D. students in agricultural economics departments (including myself in the late 1970s) were kept busy trying to determine whether Schuh was right or if he was just talking through his hat. Regardless of what each of them decided, they owe Ed Schuh a tremendous intellectual debt because he clearly and early pointed out the best type of problem for a young (or older) agricultural economist to work on: one that was clearly unresolved (there was more to be done than just dotting i's and crossing t's), and at the same time, had tremendous practical relevance.

The basic idea of this paper is that the U.S. exchange-rate policy of the postwar era had effectively taxed agricultural exports, thus, tending to diminish agricultural exports, agricultural prices, agricultural incomes, returns to farm labor, and ultimately agricultural land values. Finally, what is often neglected about the Schuh argument is that he also used the induced innovation hypothesis in combination with an overvalued exchange rate to explain partially the shift away from land-intensive cultivation practices in the United States to capital-intensive practices.

Some of the hypotheses of Schuh were immediately criticized on the basis of empirical work by Greenshields. The Greenshields paper pursued an empirical answer to the question of what effect the exchange rate had on Japanese-United States agricultural trade. He found that the exchange rate had not been a significant determinant of the level of agricultural trade between those two nations. Johnson, Grennes, and Thursby then applied a version of the Armington constant elasticity trade model to the world wheat market and found the exchange rate to be a less important determinant of trade flows and agricultural prices than foreign domestic farm policies. These papers and several others were criticized in a paper by myself and Richard Just. The basic contention was that the exchange-rate specification used in earlier papers was unduly restrictive and forced the reduced-form, exchange-rate elasticity for the prices of agricultural commodities to lie within the unit circle. This argument was heavily criticized in a comment by Grennes, Johnson, and Thursby and another comment by Reed. The Chambers and Just argument has yet to be shown to be theoretically incorrect and much of the remaining dispute centered around its empirical relevance.

Because that paper (as any paper whose title starts "A Critique of..." and then proceeds to name names is likely to) engendered some controversy, it is worthwhile to review the argument informally. The gist of the argument is that cross-price effects matter and that examinations of the effects of exchange-rate effects that either explicitly or implicitly force cross-price effects to be zero were open to empirical and analytical question. However, David Orden recently pointed out that Chambers and Just themselves committed an error of omission when they ignored the effects on income that changes in the trade balance (caused by changes in the exchange rate) could have. He has shown that even apart from cross-price effects, good reasons exist to believe that the reduced-form elasticity of an agricultural price with respect to the exchange rate is outside the unit interval. Yet another valid criticism of Chambers and Just and, indeed, of much of the literature on this subject was offered by Nishiyama and Rausser who point out that Chambers and Just ignore the effects of international currency arbitrage on agricultural prices and quantities. Simply stated their argument is that when the exchange rate between any two countries' currencies change, this sets in motion changes in other bilateral exchange rates that can have effects on agricultural exports and prices. They then proceed to model theoretically and then empirically measure these divergences. Their empirical evidence supports the practical relevance of these linkages. A closing comment on their analysis is that it is particularly informative because it explicitly accounts for the existence of price-distorting, price-support policies in the United States.

Support for the Chambers and Just (1979) contention was forthcoming in Chambers (1979) and Chambers and Just (1981) which suggested that the elasticity of the prices of soybeans and corn did lie outside the unit interval and that changes in the exchange rate were important in determining U.S. agricultural trade flows. The exchange-rate elasticities reported in those studies were, to my knowledge, the first econometric attempt to ascertain the simultaneous effect of exchange rate changes on individual agricultural prices, agricultural exports, agricultural inventories, and agricultural production. Previous modeling efforts had either relied greatly upon judgmental models or upon ones which were primarily simulation models and not sets of consistently estimated equations. For the econometric evidence with which I am familiar, Chambers and Just (1981) isolate the largest exchange rate effects on agricultural markets. There are several reasons why this has happened. First, the estimation period Chambers and Just used covered the period when the United States moved from a fixed to a flexible exchange rate. And second, the linear specification which Chambers and Just used treated the exchange rate as a separate variable in regressions that covered a period where there were other dramatic changes in agricultural export markets that were almost exactly contemporaneous with the move from fixed to flexible exchange rates. One expects large real changes to occur when there are large real structural shifts as would be associated with a movement from a fixed to a flexible exchange rate regime. This is really the essence of the Orcutt hypothesis that figured prominently in the first Chambers and Just paper (1979). However, once the exchange rate is endogenous and determined in conjunction with all prices, exchange rate changes should evince less adjustment than under a fixed exchange rate regime. The forces that ultimately determine the exchange rate are simultaneously affecting all prices. And apart from some type of block recursive structure that one might impose on agriculture it can no longer be said that the exchange rate "causes" changes in the agricultural sector.

Collins, Meyers, and Bredahl have argued that appropriate modeling of exchange rate effects requires recognizing that many countries pursue agricultural or sectoral policies that insulate domestic markets from international markets. They find that policies which tend to insulate world markets from domestic markets also mitigate the effects of exchange rate changes. Bredahl made a very similar point in an earlier paper. An easy example that illustrates is given by how producers facing a target price respond to a depreciating exchange rate. Typically, one expects a depreciating exchange rate to shift the total demand curve facing producers of a farm commodity. If cross price effects are ignored, it is easy to show that unless the supply curve is perfectly inelastic, the percentage rise in price will always be less than the percentage change in the exchange rate (this is the result that Chambers and Just show will not generally hold). However, with a target price, the effective market supply curve for all prices below the target price is perfectly inelastic. If the shift in the total demand curve caused by the exchange rate change is not large enough to force producers to face the market price, then all market price adjustment is made on the demand side and the percentage rise in the price exactly equals the percentage change in the exchange rate. However, quantity produced and consumed does not change.

Although Collins, Meyers, and Bredahl verify their hypothesis with some empirical analysis, their ideas were pursued more fully in a paper by Longmire and Morey. The Longmire and Morey paper represents the most ambitious attempt to ascertain the effects of exchange rates on agricultural trade and prices. This paper combines the arguments of Chambers and Just (1979) and Collins, Meyers, and Bredahl into a single model. Although they did not econometrically estimate the relevant elasticities, this paper (in my opinion) represents the most thorough empirical examination of this issue. One of their most important conclusions was that because of fixed loan rates, the rapid appreciation of the dollar in the early 1980s led to a significant accumulation of government inventories. The basic reasoning here is simple enough although it wasn't clearly recognized by many analysts until the Longmire and Morey paper pointed it out (see, however, Schuh pp. 5-9). An appreciating currency depresses agricultural exports and agricultural prices. As prices fall toward the loan rate more and more farmers find it profitable to forfeit their commodity under loan to the government rather than redeeming the loans. Thus, CCC inventories build up. Longmire and Morey found that as much as $2 billion of grain moved into CCC bins in 1981 and 1982 as a result of the dollar appreciation. If correct, as seems plausible, one easily sees the implications that changes in the exchange rate can have for agricultural policy.

Not all empirical studies agree, however, with the importance that Chambers and Just, and Longmire and Morey attach to changes in the exchange rate for agriculture. Batten and Belongia concluded that although the exchange rate was an important determinant of agricultural trade flows that the impact of real exchange rates on agricultural trade "...was dominated by the level of real GNP in importing nations." Although there are some obvious quibbles that could be raised with the *ad hoc* nature of the Batten and Belongia analysis, this seems to me to be a quite plausible result. That is not to say that I agree with their claims as to the magnitude of how much variation in agricultural exports can be explained by changes in real GNP in importing nations. But just as clearly in the early 1980s many important markets for U.S. farm products were experiencing either severe debt problems or were in the grips of a world-wide recession. So, the importance of this structural fact should not be overlooked. The obvious policy conclusion to be drawn from this is that to encourage agricultural export expansion it is probably more important in the long run to foster economic growth abroad than to try to influence exchange rates artificially.

The other major contribution of the Batten and Belongia paper was to emphasize the role that real and not nominal exchange rates play in determining trade flows. Although I think this fact was well appreciated by many since it is the implicit assumption that underlines many of the studies of the effect of macroeconomic phenomena on agriculture (see, for example, the extended discussion of this point in Chambers (1985)), it was a point well worth making.

So far, the discussion has centered on studies of the effects of exchange rates on U.S. agricultural trade. However, long before the effect of the exchange rate became an important topic for American agricultural economists, it was a well-understood phenomenon for Australian agricultural economists. Although I will not attempt to survey that literature, I would encourage anyone interested in the general topic to peruse the Australian Journal of Agricultural Economics.

There are also studies of the Canadian exchange-rate experience that American readers would benefit from reading. One of the earliest empirical studies to incorporate the exchange rate in an agricultural trade model was Meilke and de Gorter. Recently, Meilke and Coleman analyzed the effects of the exchange rate on the Canadian red-meat sector. They find that exchange rate fluctuation between the Canadian and U.S. dollars have different effects on the differing segments of the Canadian red-meat industry. For instance, segments of the industry which predominantly export to the United States, but which do not rely upon imports from the United States (like cow-calf operators) gain from a depreciation of the Canadian dollar. Segments of the industry which rely upon imports from the United States, but which continue to export to the United States will see gains made from a depreciation eroded by the losses they face in purchasing now more expensive U.S. inputs. This raises an interesting theoretical issue which has not been adequately addressed in the agricultural economics literature. And that is what is the ultimate effect of changes in exchange rates when inputs are traded internationally.

Although a number of empirical studies have been done on the effects of exchange rates upon agriculture, very little attention has been paid to the empirically important issue of which exchange rate index is appropriate for empirical trade models. It is not an exaggeration to say that almost all of the above studies used a different exchange rate variable for their empirical analysis. Recently, Dutton and Grennes have addressed the issue of the appropriate exchange rate index to use in agricultural trade models. Their discussion is based on the general theory of index numbers. Their approach seems to have been one of looking at what can be inferred about the general structure of an index number from economic theory and then using these general inferences to write down plausible index numbers. The issue of constructing a plausible exchange rate index is not a trivial extension of the usual theory of index numbers in my eyes. Dutton and Grennes seem to favor deriving separate exchange rate indexes for imported commodities and for exported commodities. This is quite in line, for example, with the practice of USDA in computing trade-weighted exchange rates for various agricultural commodities. However, this leads Dutton and Grennes to suggest basing exchange-rate indexes for export commodities upon the "product function" and exchange-rate indexes for imported commodities upon the national expenditure function. While this approach seems reasonable, I would suggest that another approach which leads to different answers should be given some consideration, i.e., to base exchange rate indexes on the indirect trade utility function which considers both preferences and the technology that exists at any point in time. There are good reasons to think that this approach may be more meaningful--chief amongst them is the fact that both preferences and the state of technology interact simultaneously to determine whether a country is a net importer or a net exporter. Thus, basing an index upon either the product function or the expenditure function ignores half of the problem. And when preferences and technology differ significantly across nations, these differences should be taken into account in constructing such indexes. To illustrate the approach that I am suggesting consider the indirect trade utility function defined by:

$$H(p,x) = V(p, R(p,x))$$

where

$$V(p,z) = \text{Max}\{u(y):py \leq z\}, \text{ and}$$
$$R(p,x) = \text{Max}\{py: (x,y) \in T\}.$$

Here p is an n dimensional vector of strictly positive output prices, x is an m dimensional vector of inputs representing a country's resource endowment, u(y) is a concave utility function, y is an n dimensional vector of outputs, and T is a compact set representing technology available to society. Now denote p^1 as the subvector of traded goods prices and p^2 as the subvector of nontraded goods prices. Finally, consider the implicit function defined by the equality:

$$H(p,x) = u.$$

If we solve this equality for a single input in terms of the other inputs, the output prices and the level of utility we achieve an analogue of the expenditure function which I shall term the trade expenditure function and denote as:

$$x_m = E(p,u,\hat{x})$$

where $\hat{x}$ is the complement of x_m in x. E( ) gives the level of the m-th input required to attain utility equal to u with a given resource endowment for the other inputs and the fixed vector of prices p. A natural, real price index in this context is to consider the level of x_m required to purchase the same level of u at differing output prices but at the same endowment of $\hat{x}$. Now for there to exist a subindex that only depends upon the prices and quantities of the traded goods, it is necessary that the indirect trade utility function has p^1 separable from p^2. This implies that the ratio

$$\frac{\partial H(p,x)/\partial p_i}{\partial H(p,x)/\partial p_j}$$

be independent of the elements of p^2 when both p_i and p_j belong to p^1. But direct differentiation reveals that

$$\partial H(p,x)/\partial p_i = \partial V(p,m)/\partial p_i + (\partial V(p,m)/\partial m)(\partial R(p,x)/\partial p_i).$$

Hence, imposing separability upon the structure of H(p,x) entails making structural assumptions about both the shape of preferences and the form of the technology.

The results reported by Dutton and Grennes are extremely interesting, however. In general, they find that the degree of appreciation or depreciation varies in a nonnegligible fashion across the differing exchange-rate indexes. Some indexes, in particular, tend to either understate or overstate the degree of appreciation or depreciation. The interesting point here is that these results could potentially lay to rest much of the controversy over the relative magnitude of exchange rate effects. If, as appears to be true, there is a nontrivial measurement problem involved, then the wide range of elasticities reported in the literature could well be consistent with one another. For example, the relatively large elasticities reported by Chambers and Just might be attributed to the inability of the exchange rate index that they used to adequately measure the true degree of depreciation that occurred in the 1970s. The large effect was there; it just may have been the case that the exchange rate depreciation was more significant than that implied by the Chambers and Just exchange rate index.

MACROECONOMICS AND AGRICULTURE

An immediate outgrowth of the literature on the effects of exchange rates on agricultural trade was an increased sensitivity on the part of agricultural economists to the role that forces previously seen as exogenous to the agricultural sector played in determining outcomes in agricultural markets. Just as Richard Nixon once remarked, "We're all Keynesians now," all prominent agricultural economists promptly became experts on the effects of macroeconomic phenomena on agriculture. And at first, an agricultural economist did not really have to know anything about modern macroeconomics to be considered a macroeconomist by his colleagues in agricultural economics. All he had to do, quite literally, was to include a macroeconomic variable in an agricultural model and he qualified.

Since agricultural economics had always been very oriented toward microeconomics, it was not surprising, therefore, that in this climate, even some usually very reasonable people did and said some very unreasonable things. A prime example was the debate that raged about whether or not inflation was neutral with regard to agriculture. That debate gave new meaning to the words *ad hoc*. The situation reached such a point where an individual could construct a general equilibrium production model, link it to a portfolio-balance model of financial asset markets and then be accused of conducting macroeconomic research, when all along the individual had been convinced that he had adopted a microeconomic approach to the problem. Fortunately, however, the situation has improved somewhat in recent years. Just as it took agricultural economics almost ten years to train a group of serious international agricultural economists, it has taken a number of years to gear up in the macroeconomics area.

Shei represents an early study looking at the effects of macroeconomic variables on the agricultural economy. He constructed a general equilibrium model that included agriculture as a separate sector in a fixed exchange rate world and then used the resulting model to simulate the effects of various monetary policy alternatives on agriculture. He concluded that agricultural and nonagricultural prices were differentially affected by monetary policies with agricultural prices adjusting more rapidly than nonagricultural prices. This work was later followed by a companion paper by Shei and Thompson (chapter 4) that focused on the relationship between inflation and agriculture while combining both structuralist and monetarist characteristics.

Chambers (1979) constructed an econometric model of the linkage between financial and agricultural markets that emphasized the role of the exchange rate in transmitting monetary phenomena to the agricultural economy. This work was followed by Chambers and Just (1982) which examined the implications of changes in monetary policy for U.S. agriculture and concluded that the potential effects were quite large. These two studies represent the first attempt that I am aware of to incorporate both macroeconomic variables and agricultural variables in a simultaneous equations model with feedback effects from agriculture to the macroeconomy.

Belongia and King developed a model relating rates of change in the food consumer price index to changes in the level of the money supply and several other variables using a modification of the quantity theory of money. Their empirical research, however, does not support the contention that changes in monetary policy have significant effects on real food prices.

Several studies conducted "causality" tests to determine if agricultural prices (both real and nominal) were "caused" by the level of the money supply or other liquidity measures (Barnett; Barnett, Bessler and Thompson; Bessler; and Chambers (1981)). Some evidence of a causal relationship between monetary variables and some agricultural prices were detected. This is particularly interesting because each study used differing time periods and differing definitions of liquidity. Barnett used monthly data, for example, while Chambers (1981) used annual data for the period 1892-1952. The Chambers (1981) study is particularly interesting because it isolated support for the contention that money caused changes in wheat exports, i.e., a nominal variable caused real changes. Although such studies are by and large *ad hoc* and, therefore, should be interpreted with extreme caution, they are interesting in that a detectable statistical relationship between the money supply and agricultural price variables for widely varying time periods and market structures was found.

The linchpin of many of these studies and of most remaining studies of the effects of macroeconomic and particularly monetary phenomena on agriculture has been the presumption of some type of price fixity in agriculture or in the general economy. This is explicit in Shei's model where the exchange rate is fixed and some nonagricultural goods prices are fixed. It is implicit in the Chambers (1979) and the Chambers and Just (1982) studies which consider real agricultural prices, but only have explicit equations sufficient to identify the agricultural prices alone, i.e., there is no separate equation for the price deflator. The importance of the fixity of some prices within the economy in these investigations was recognized in Chambers (1984) and has become the focal point of research for a group of economists at Berkeley.

The idea that some prices adjust as a result of market forces while others do not is not really new. It dates at least to the time of Hicks' fixprice-flexprice hypothesis. And it is not even necessary that prices be fixed; all that is really necessary is that there be differential rates of adjustment. The reason is that in a perfectly frictionless economy, money simply doesn't matter to real decisions. However, in an economy where there is some friction in adjustment, monetary changes can have real effects. For example, suppose agricultural prices are flexible and nonagricultural prices are not. Then inflating the money supply will make agricultural prices rise while nonagricultural prices will not. However, this means that real agricultural prices have risen and this will have real effects on the economy in terms of increased agricultural production, decreased agricultural consumption, and likely decreased stockholding.

The theoretical implications of changes in monetary policy for agricultural exports, agricultural prices, agricultural incomes, returns to differing agricultural factors of production, and agricultural production were worked out in Chambers (1984) for a fixed-flex price economy. Not surprisingly, many of the theoretical implications were of the "it depends..." type in that study. The same study also carried out a vector autoregression analysis of a subset of these relationships. The empirical analysis suggested that innovations in monetary policy, in fact, did cause real changes in the agricultural economy with restrictive monetary policy lowering agricultural prices relative to the prices on nonagricultural commodities, discouraging net agricultural exports, depressing farm incomes. The analysis, however, also tended to suggest that events arising in agriculture accounted for more of the variation in the agricultural variables than the monetary policy.

Perhaps the most thorough analysis of the effects of monetary policy on agriculture in the presence of fix-flex prices has been carried out by Gordon Rausser and Kostas Stamoulis and their colleagues at the University of California, Berkeley. Their work focuses on the overshooting phenomena which was originally formulated in terms of exchange rate models by Dornbusch. The Stamoulis and Rausser paper that follows is explicitly devoted to the overshooting phenomenon so I only touch on its basic content here. In a world where some prices are fixed and other prices are flexible, short-run adjustment in the flexible prices can overshoot the long-run adjustment in the flexible prices that would be caused by, say, changes in monetary policy. Thus, flexible prices adjust more to monetary policy in the short run than they would in the long run.

Since the phenomenon of overshooting has and continues to receive much attention, I will divert my attention for a moment and give the reader a simple example that I think illustrates the essence of the overshooting problem. Consider the following model: let the price that adjusts relatively slowly be denoted p_1 and the price of the commodity that adjusts more rapidly be denoted p_2. Assume that the intertemporal behavior of p_1 is depicted by the following convergent equation of motion:

$$dp_1/dt = g(p_1,\alpha)$$

while the demand and supply of the second commodity is given by

$$D(p_1, p_2, \alpha), \text{ and}$$
$$s(p_2).$$

It is assumed that the demand curve is downward sloping in the own price and that the supply curve is upward sloping. Moreover, assume that D_1 and $g\alpha$ are both positive while $D\alpha$ is negative. The original equilibrium is depicted in Figure 1, panels a and b. The steady-state price for p_1 is p_1^0 while the equilibrium price for p_2 is p_2^0. Now suppose that an exogenous shock occurs and α rises; this tends to shift the demand for the second commodity back to the curve labeled D_1^0 which is demand evaluated at the long-run equilibrium p_1^0 and the new α. In the very short run, p_1 will not be able to adjust because adjustment takes time for sticky price commodities. Therefore, the short-run equilibrium is at price p_2^s. But in the long-run, the change in α also affects the long-run value of p_1 as is illustrated by g shifting to the right in panel a. The new long-run equilibrium occurs at price p_1^*. But as p_1 rises to this level the demand for the second commodity starts to shift back towards its original position ending finally at its long-run position which is labeled D^* in panel b. Consequently, the long-run equilibrium value of p_2 is p_2^*, and we see that the short-run adjustment did, indeed, overshoot the long-run adjustment.

Now this example, although very stylized and simple, is important for several reasons because it illustrates an essential part of the overshooting problem and because it also points out some interesting results. First, as is clear from the way the assumptions are structured, overshooting need not always occur. If the assumptions are changed, overshooting may not occur. Secondly, there can be instances where short-run effects could be in the exact opposite direction of long-run effects. Suppose, for example, that all of the assumptions remain the same except for the fact that there is a very strong substitute relationship between the two commodities, so strong, in fact, that the rise in p_1 in the long-run is enough to compensate for the change in α and shift the long-run demand curve to D^{**} and not D^*. As illustrated, the long-run value of p_2 rises, not falls. Finally, it is clear that even in the absence of long-run effects, the existence of sticky prices can lead to important short-run, real effects. This would occur, for example, if the long-run rise in p_1 caused by the change in α were just sufficient to bring the demand curve back to its original position. Here there are real, short-run effects and production actually changes. But in the long-run, there is no effect. Such a situation is a stylized representation of the long-run neutrality of the second commodity's market to changes in α.

FIGURE 1
OVERSHOOTING

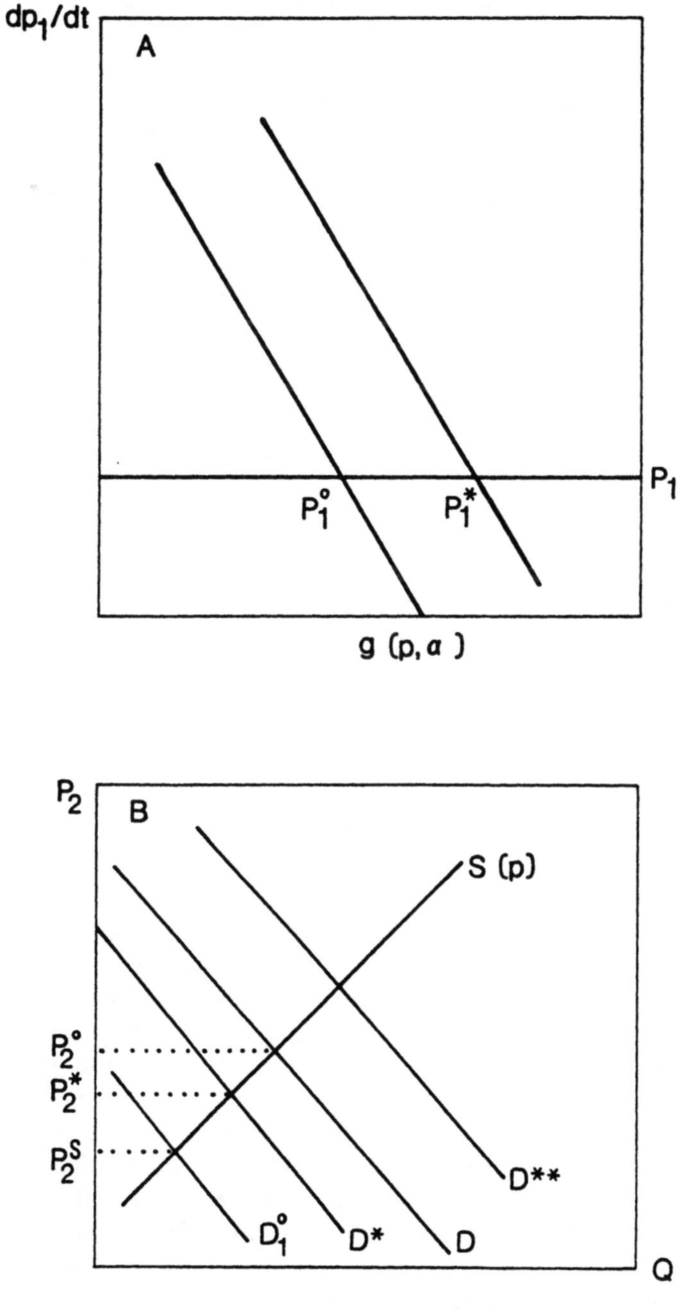

A second major contribution of Rausser, et. al., is that they have paid close attention to modelling realistically the effects of agricultural policy actions on the adjustments to changes in monetary policy and other macroeconomic phenomena. They argue in essence that loose monetary policy (loosely speaking) tends to subsidize agriculture in the short run while tight monetary policy (still loosely speaking) tends to tax agriculture in the short run. When these tax components are combined with the structural rigidities of agricultural policy in nominal terms (loan rates fixed in nominal and not relative terms) then real changes in the agricultural economy are causes which tend to exacerbate the already unstable nature of the agricultural economy. While I cannot take the time here to analyze the contribution of Rausser and his colleagues in general, I can say it is significant, and I would refer the interested reader to Rausser and Rausser, Chalfant, Love, and Stamoulis and to the paper of Stamoulis and Rausser which follows as representative samplings.

Recently, Orden has used VAR methods to analyze the effects of the money supply, the exchange rate, and the interest rate on agricultural exports and prices received by farmers. His VAR model is similar to that included in the Chambers (1984) paper except that he broadens the list of monetary instruments. On the basis of his analysis, he is able to conclude that these variables played a significant role in determining agricultural exports and relative agricultural prices. Orden further concludes that "...monetary policy is a powerful determinant of relative agricultural prices...when monetary shocks are relatively large. Autonomous interest rate and exchange rate shocks of relatively smaller magnitudes have a greater effect on the value of agricultural exports than shocks to the money supply variable...but are not subject to control by macroeconomic policymakers."

I have taken Dave's model and reestimated it using a slightly different specification that includes an export-weighted support rate in the causal chain. Also my results (figure 3) are not strictly comparable because I have not used the exact same sample period and the exact same variable descriptions. However, the results are interesting for two reasons: one, they pursue an important point that has been raised by Longmire and Morey, i.e., the structure of domestic policy programs can have important implications for the effects of monetary policy; and two, these results illustrate the sensitivity of VAR models to changes in data definitions and causal orderings. As you can see, the results that I am presenting (figure 3) are somewhat different from what Dave has found (figure 2).

FIGURE 2

MONETARY POLICY AND RELATIVE AGRICULTURAL PRICES

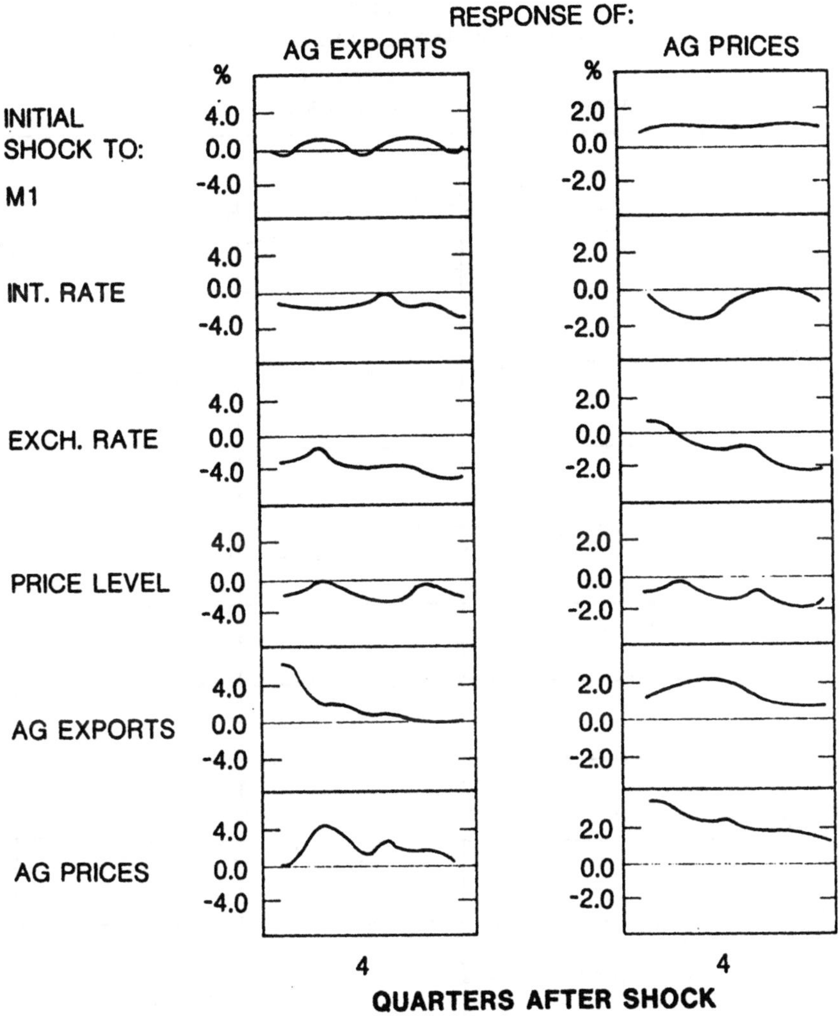

Reprinted by permission of Elsevier Science Publishing Co., Inc. "Agriculture, Trade and Macroeconomics: The U.S. Case" David Orden, Journal of Policy Modeling, 8(1): 27-51(1986) © Society for Policy Modeling, 1986.

FIGURE 3

LOAN RATES AND RELATIVE AGRICULTURAL PRICES

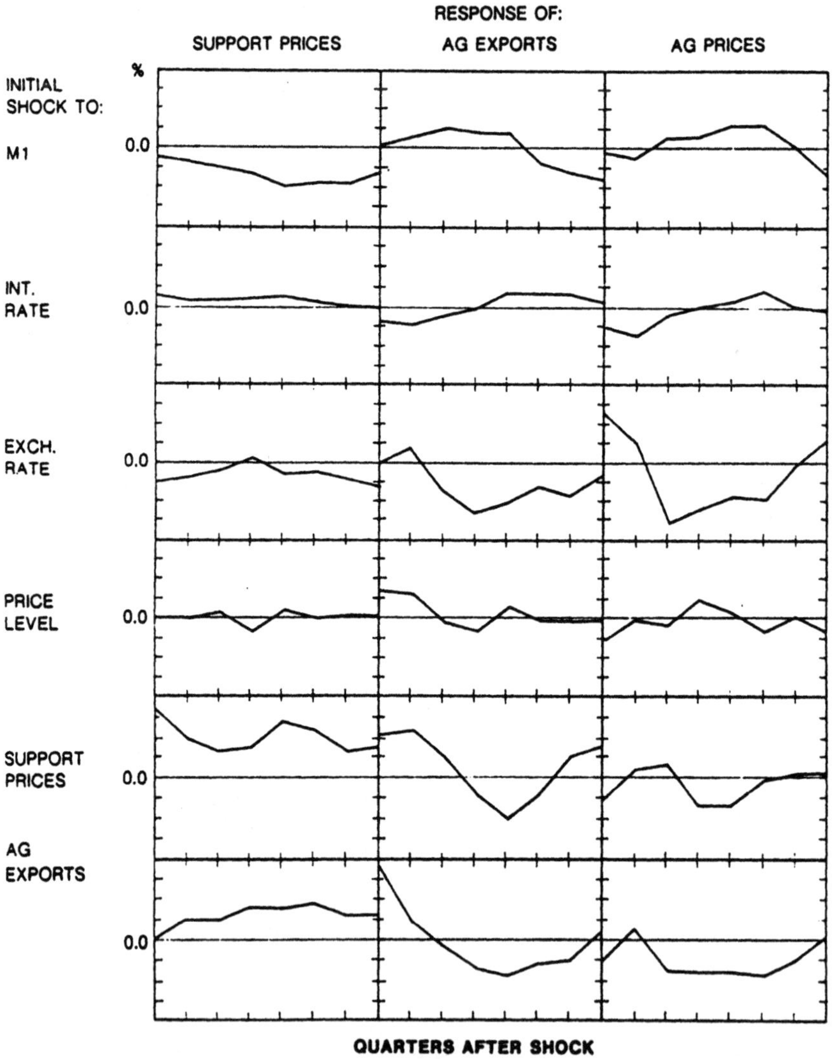

Most of the forecast error variance in the agricultural price variable is attributable to changes in the exchange rate in the very short run with the effect diminishing over time while in the short run, monetary policy has relatively little effect on agricultural prices and a much more important effect after 12 months have elapsed. In the long run, i.e., 12 months, variables usually seen as exogenous to agriculture (the exchange rate, the money supply, and support prices) account for about 65 percent of the forecast error variance while factors usually seen as jointly dependent with agricultural prices (agricultural exports and prices received by farmers) only account for a little over 26 percent. While I am convinced that this model could be slightly changed and rerun to obtain vastly different results, I do find the results striking enough to suggest that models of macroeconomic linkages to agriculture should pay careful attention to structuring realistic models of agricultural policy.

REFERENCES

Barnett, R. The Relation between Domestic and International Liquidity and Nominal Agricultural Prices. MS thesis, Purdue University, 1980.

Barnett, R., D. Bessler, and R. L. Thompson. "Agricultural Prices in 1970s and the Quantity of Money." Presented at AAEA annual meeting, Clemson, SC, 26-29 July 1981.

Batten, D. and M. T. Belongia. "The Recent Decline in Agricultural Exports: Is the Exchange Rate the Culprit?" Federal Reserve Bank of St. Louis Review. 66(1984), October:5-14.

Belongia, M. T. and R. King. "A Monetary Analysis of Food Price Determination." American Journal of Agricultural Economics. 65(1983): 131-5.

Bessler, D. A. "Relative Prices and Money: A Vector Autoregression on Brazilian Data" American Journal of Agricultural Economics. 66(1984): 13-19.

Bredahl, M. E. "Effects of Currency Adjustments Given Free Trade, Trade Restrictions and Cross Commodity Effects." Dept. Agr. Econ. Staff Paper P76-35, University of Minnesota, St. Paul, Nov. 1976.

Chambers, R. G. An Econometric Investigation of the Effect of Exchange Rate and Monetary Fluctuation on US Agriculture. Ph.D. thesis, University of California, Berkeley, 1979.

_______. "Credit Constraints, Interest Rates, and Agricultural Prices." American Journal of Agricultural Economics. 67(1985): 390-395.

_______. "Interrelationships between Monetary Instruments and Agricultural Commodity Trade." American Journal of Agricultural Economics. 63(1981): 934-41.

_______. "Agricultural and Financial Market Interdependence in the Short Run." American Journal of Agricultural Economics. 66(1984): 12-24.

Chambers, R. G. and R. E. Just. "A Critique of Exchange Rate Treatment in Agricultural Trade Models." American Journal of Agricultural Economics. 61(1979): 249-57.

_______. "Effects of Exchange Rate Changes on US Agriculture: A Dynamic Analysis." American Journal of Agricultural Economics. 63(1981): 32-46.

Chambers, R. G. and R. E. Just. "Investigation of the Effects of Monetary Factors on US Agriculture." Journal of Monetary Economics. 9(1982): 235-47.

Collins, K.J., W. H. Meyers, and M. E. Bredahl. "Multiple Exchange Rate Changes and US Agricultural Commodity Prices." American Journal of Agricultural Economics. 62(1980): 656-65.

Dornbusch, R. "Expectations and Exchange Rate Dynamics." Journal of Political Economy. 84(1976): 1161-1175.

Greenshields, B. F. "Changes in Exchange Rates: Impact on US Grain and Soybean Exports to Japan." FAER 364. Economic Research Service, U.S. Department of Agriculture, Washington, D. C. July 1974.

Grennes, T., P. R. Johnson, and M. C. Thursby. "A Critique of Exchange Rate Treatment in Agricultural Trade Models: Comment." American Journal of Agricultural Economics. 62(1980): 249-252.

Johnson, P. R., T. Grennes, and M. Thursby. "Devaluation, Foreign Trade Controls, and Domestic Wheat Prices." American Journal of Agricultural Economics. 59(1979): 619-27.

Longmire, J. and A. Morey. "Strong Dollar Dampens Demand for US Farm Products." FAER 193. Economic Research Service, U. S. Department of Agriculture. Washington, D.C., 1983.

Meilke, K. and H. de Gorter. "A Quarterly Econometric Model of the North American Grain Industry," presented to Economics Branch, Agriculture Canada, Ottawa, 27 Apr. 1977.

Nishiyama, Y. and G. C. Rausser. "Multiple Effects of Exchange Rates on Import Demand: The Case of U.S. Agricultural Trade with Japan." Unpublished working paper, University of California, Berkeley, June 1986.

Orden, D. "A Critique of Exchange Rate Treatment in Agricultural Trade Models: Comment." American Journal of Agricultural Economics 68(1986): 990-993.

_______. "Agriculture, Trade, and Macroeconomics: The US Case." Journal of Policy Modelling. 8(1986): 27-51.

Rausser, G. C. "Macroeconomics and US Agricultural Policy." Studies in Economic Policy, American Enterprise Institute, 1986.

Rausser, G. C., J. A. Chalfant, H. A. Love, and K. Stamoulis. "Macroeconomic Linkages, Taxes, and Subsidies in the US Agricultural Sector," American Journal of Agricultural Economic 68(1986): 399-412.

Reed, M. "A Critique of Exchange Rate Treatment in Agricultural Trade Models: Comment." American Journal of Agricultural Economics. 62(1980): 253-254.

Schuh, G. E. "The Exchange Rate and US Agriculture," American Journal of Agricultural Economics. 56(1974): 1-13.

Shei, Shun-yi. The Exchange Rate and United States Agricultural Product Markets: A General Approach. Ph.D. thesis, Purdue University, 1978.

PART I

EXCHANGE RATES: SOME VIEWS FROM INTERNATIONAL FINANCE

2
Explaining the Demand for Dollars: International Rates of Return, and the Expectations of Chartists and Fundamentalists

Jeffrey A. Frankel and Kenneth A. Froot

The careening path of the dollar in recent years has shattered more than historical records and the financial health of some speculators. It has also helped to shatter faith in economists' models of the determination of exchange rates. We have understood for some time that under conditions of high international capital mobility, currency values will move sharply and unexpectedly in response to new information. Even so, actual movements of exchange rates have been puzzling in two major respects.

First, the proportion of exchange-rate changes that we are able to predict seems to be, not just low, but zero. According to rational expectations theory we should be able to use our models to predict that proportion of exchange rate changes that are correctly predicted by exchange market participants. Yet neither models based on economic fundamentals, nor simple time series models, nor the forecasts of market participants as reflected in the forward discount or in survey data, seem able to predict better than the lagged spot rate. Second, the proportion of exchange rate movements that can be explained even after the fact, using contemporaneous macroeconomic variables, is disturbingly low.

Jeffrey A. Frankel, Department of Economics, University of California-Berkeley, Berkeley, CA. and Kenneth A. Froot, Sloan School of Management, Massachusetts Institute of Technology--Cambridge, MA.

FUNDAMENTALS, BUBBLES, AND TESTS OF RATIONAL EXPECTATIONS

Most of the models of exchange rate determination that were developed after 1973 are driven by countries' supplies of assets: supplies of money alone in the case of the monetary models, and supplies of bonds and other assets as well in the case of the portfolio-balance models.[1] But observed supplies of dollar assets versus other currencies are no help in explaining the 1981-85 appreciation of the dollar. The supply of U.S. assets was increasing rapidly, as measured by the federal government deficit (or the money supply). At the same time, the stock of net claims against foreigners has been decreasing rapidly as measured by the current account deficit.

There is general agreement that the 1981-85 appreciation of the dollar was attributable to an increase in the demand for dollars on the part of investors worldwide. There is much less agreement as to the cause of that change in demand. Four hypotheses have been commonly proposed as to why investors found U.S. assets more attractive in the early 1980s. The first, which might be termed "monetarist," is that there was a decline in the rate of expected inflation and depreciation after 1980 because of a reduced rate of money growth.[2] The second is that there was an increase in the interest differential relative to the expected rates of inflation or depreciation; this is the "overshooting" explanation.[3] The third is that there was a self-confirming fall in the expected rate of dollar depreciation; this is the "speculative bubble" hypothesis. Each of these three attributes the increase in demand for assets to an increase in the expected rate of return, variously defined. The fourth, the "safe haven hypothesis" is different; it attributes the shift in demand to an increase in the perceived safety of U.S. assets relative to other countries' assets.

In the first half of the paper we consider briefly each of these four explanations by means of the data on expected returns for the period reported in Table 2.1. Of the three that depend on economic fundamentals--the monetarist, overshooting, and safe haven hypotheses--we argue that only the second is capable of explaining the large real appreciation of the dollar from 1981 to 1985 and its subsequent depreciation. But even the overshooting model seems unable to explain entirely the path taken by the dollar and particularly the last 20 percent of appreciation preceding the February 1985 peak and subsequent rapid decline.

TABLE 2.1
Rate of Return Differentials on US Assets Relative to Trading Partners
(% per annum)

Expected Inflation Differential	Years 1976-78	1979-80	1981-82	1983-84	1985-86
1 One-Year lag	-1.01	3.54	0.88	-0.35	0.06
2 Three-Year distributed lag	-1.96	2.7	1.89	-0.18	-0.16
3 DRI three-year forecast**	NA	2.20	0.96	0.23	0.15
4 OECD two-year forecast***	1.42	2.24	0.62	0.61	0.78
5 American Express Survey+	NA	NA	4.11	2.68	-0.16
Nominal Interest Differential					
6 One-Year interest differential*	-0.48	2.29	3.00	1.73	1.15
7 One-Year forward discount****	0.18	2.57	3.34	1.85	0.21
8 Ten-Year interest differential	-0.50	0.56	1.91	2.47	2.92
Real Interest Differential					
9 One-Year (6-1)	0.53	-1.24	2.12	2.09	1.08
10 Ten-Year w/ distributed lag (8-2)	1.47	-2.15	0.02	2.64	3.08
11 Ten-Year w/ DRI forecast (8-3)	NA	-1.64	0.95	2.24	2.77
12 Ten-Year w/ OECD forecast (8-4)	-1.92	-1.68	1.29	1.86	3.12
Expected Depreciation From Surveys++					
13 Economist 3 Month	NA	NA	12.99	10.10	1.50
14 Economist 6 Month	NA	NA	10.62	10.78	4.99
15 Amex 6 Month	2.08	NA	9.54	7.21	1.39
16 Economist 12 Month	NA	NA	8.57	8.60	5.41
17 Amex 12 Month	0.61	NA	6.67	6.99	3.75
18 (7/15)	NA	NA	0.31	0.17	0.04

*Calculated as ln(1+i), 1985 contains data through June, rates for Japan not available 1976-77. **Averages of various forecast dates, through March 1985. ***OECD forecasts available during 1976-78 only for 12/78, during 1985 for June 1985. ****Available at 11 survey dates. +Available at 11 survey dates only for US, UK, WG, and at 4 survey dates (76-78) for France. ++See Frankel and Froot (1985) for an explanation of the survey data, including the dates on which surveys were conducted. Expected depreciation uses GNP-weights for UK, FR, WG and JA. ****Includes data through February 1986.

Sources: IMF International Financial Statistics, DRI FACS financial data base and forecasts, OECD Economic Outlook, Capital International Perspective, AMEX Bank Review, and Economist Financial Review.

Note: Differential calculated as US - foreign, where foreign is a GNP weighted average of UK, FR, WG, and JA unless otherwise specified.

In the second half of this paper we propose the outlines of a model of a speculative bubble that is not constrained by the assumption of rational expectations. The model features three classes of actors: fundamentalists, chartists and portfolio managers. None of the three acts utterly irrationally; each performs the specific task assigned him in a reasonable, realistic way. Fundamentalists think of the exchange rate according to a model--say, the overshooting model for the sake of concreteness--that would be exactly correct *if there were no chartists in the world.* Chartists do not have fundamentals such as the long-run equilibrium rate in their information set; instead they use autoregressive models--say, simple extrapolation for the sake of concreteness--that have only the time series of the exchange rate itself in the information set. Finally portfolio managers, the actors who actually buy and sell foreign assets, form their expectations as a weighted average of the predictions of the fundamentalists and chartists. The portfolio managers update the weights over time in a rational Bayesian manner according to whether the fundamentalists or the chartists have recently been doing a better job of forecasting. Thus each of the three is acting rationally subject to certain constraints. Yet the model departs from the reigning orthodoxy in that the agents could do better, in expected value terms, if they knew the complete model. When the bubble takes off, agents violate rational expectations in the sense that they learn about the model more slowly than they change it. Furthermore, the model may be unstable in the neighborhood of the fundamentals equilibrium, but stable around a value for the dollar that is far from that equilibrium.

Part 1 establishes the shortcomings of the conventional approaches, including rational expectations, to accord fully with simple empirical facts of the 1981-85 period. Part 2 elaborates the distinction between chartists and fundamentalists and offers some evidence from expectations survey data that respondents seem to form very short-term expectations more like chartists and more long-term expectations like fundamentalists. Part 3 describes the model in more detail and shows how it can work to explain the 1980-85 path of the dollar.

STANDARD EXPLANATIONS OF THE 1981-1985 APPRECIATION OF THE DOLLAR BASED ON RATES OF RETURN

We begin with the simplest view of how the demand for dollars depends on rates of return, the model associated with the monetarists. In this model there are three equivalent ways of measuring the rate at which the value of the dollar is expected to change in the future relative to foreign currencies: the expected inflation differential, the expected rate of depreciation, and the nominal interest differential. The first two variables are equal if purchasing power parity holds: the goods of different countries are essentially perfect substitutes in consumers' utility functions, and barriers to instantaneous adjustment in goods markets are low. The second and third variables are equal if uncovered interest parity holds: the assets of different countries are essentially perfect substitutes in investors' portfolios, and barriers to instantaneous adjustment in asset markets are low.

At any point in the late 1970s, the U.S. dollar was expected to lose value against foreign currencies, the mark and the yen in particular, whether the expected rate of change was thought of as the expected inflation differential, the expected rate of nominal depreciation, or the nominal interest differential. In response, investors, seeking to protect themselves against expected capital losses, had a relatively low demand for dollars and high demand for marks and yen. When a firm anti-inflationary U.S. monetary policy began to take hold in 1980, investors' expectations that the dollar would lose value began to diminish rapidly. This would account for an increase in the demand for dollars and the large appreciation of the dollar in the early 1980s.

There is no single accepted way of measuring inflation expectations. The first five rows of Table 2.1 report five measures of expected inflation that are available for the United States as well as four trading partners (France, Japan, the United Kingdom, and West Germany.) The five measures are the actual inflation rate over the preceding year, a distributed lag over the preceding three years, forecasts by Data Resources, Inc., at a three-year horizon, forecasts by the OECD at a two-year horizon, and results of a survey conducted by American Express of active participants in foreign exchange markets at a one-year horizon. By the available measures, expected inflation in the U.S. by 1979-80 had climbed to a level 2-3 points above the weighted average of trading partners. The differential declined rapidly thereafter, reaching approximately zero by 1985. Thus the expected inflation numbers appear to support the first of the three explanations of the dollar appreciation listed above.

The problem is that the decline in the expected inflation differential was not at all matched by developments in other concepts of the expected rate of change of relative currency values. Directly measuring expected changes in the exchange rate is more difficult than measuring expected changes in the price level, because the former is much more volatile than the latter. A new data set is applied to this task below. But first we look at interest rate differentials.

Row 6 in Table 2.1 reports the differential in one-year nominal interest rates between the United States and the weighted average of four trading partners. Row 7 reports the one-year forward discount; the two series should be identical if covered interest parity holds. The numbers show that by 1981-82 the short-term interest differential had reached a level of 3 per cent. Thus the real interest differential, reported in row 9, rose from -1 per cent in 1979-80 to +2 percent in 1981-82. The short-term interest differential, nominal or real peaked in 1982. However, the long-term real interest differential, which rose by 2-3 points from 1979-80 to 1981-82, depending on the measure of expected inflation used, continued to rise over the next three years. In early 1985 it stood at about 3 points by any of the three measures (up from about -2 points in 1979-80).

The increase in the real interest differential offers the explanation needed for an increase in the real value of the dollar. An increase in the nominal interest differential, if it were not offset by an increase in expected inflation or expected depreciation of the currency, would make domestic assets more attractive than foreign assets. The increased demand for domestic assets causes the dollar to appreciate until investors are happy with their holdings. If the dollar is perceived as having appreciated above its long-run equilibrium, there will be an expectation of future depreciation. The short-run equilibrium will occur where the expected future depreciation is sufficient in investors' minds to offset the interest differential.

This much is familiar from the Dornbusch (1976) overshooting model. One reason for looking at the long-term differential, rather than the short-term differential that he used, is as follows.[4] The return of the exchange rate to its long-run equilibrium value could be slow and irregular. If we want to choose a length of time long enough to be confident of having reached long-run equilibrium, 10 years might be necessary. Assume that the 10-year nominal interest differential measures the 10-year expected rate of change of the nominal exchange rate. Then the 10-year real interest differential measures the 10-year expected rate of change of the real exchange rate. With our argument that 10 years is long enough for the real exchange rate to be at its equilibrium value, it follows that the currently measured 10-year (per annum) real interest differential (multiplied by ten) tells us how far from long-run equilibrium investors consider the current real exchange rate to be. Following this logic, as of early 1985 the long-term real interest differential could "explain" a real "over-valuation" of the dollar of about 30 percent relative to its perceived long-run equilibrium and could explain a real appreciation of about 50 percent relative to 1979-80.

The foregoing calculations are rather crude, and in particular are very sensitive to the term of maturity chosen. Several points can be made in defense of the approach. First, it is supported by several regression studies.[5] Furthermore, the increases in the real interest differential and in the real value of the dollar are the results that the standard macroeconomic theory of high international capital mobility predicts will result from a fiscal expansion such as that undertaken in the United States between 1981 and 1985, that is, a fiscal expansion not accommodated by either a monetary expansion or an offsetting increase in private saving. Finally, the large depreciation of the dollar in late 1985 and early 1986, as the U.S. Congress took steps to bring the fiscal deficit under control and the Federal Reserve allowed real interest rates to fall, fits the theory well. However, as always with exchange rate theories, there are problems if one tries to fit the data on as finely as a monthly basis. In particular, the long-term real interest differential was already declining during the second half of 1984, even though the dollar continued to appreciate rapidly until February 1985. The fiscal contraction did not begin until the Gramm-Rudman budget reduction bill was passed in December 1985, or at the earliest when the Congress voted to slow the future rate of growth of military spending in mid-1985. The final stages of the dollar's ascent appear unexplained.

An alternative fundamentals explanation sometimes given for the 1981-85 appreciation of the dollar is the safe-haven hypothesis: a world wide increase in investors' demand for U.S. assets in response to a perceived decrease in the risk of assets held in the United States relative to those held elsewhere. Such a portfolio shift by itself would be inconsistent with the increase in the interest rate differential observed in Table 2.1. But the argument runs that a common set of developments--the improved treatment of investment in the 1981 tax bill and the generally improved climate for business under the Reagan Administration--is responsible for both the 1983-84 investment boom (after the investment slump of 1981-82) and the safe-haven portfolio shift, and that the former had an upward effect on real interest rates that dominated any downward effect of the latter. We will be offering some evidence against the safe-haven hypothesis in section 1.3 below. We will then turn from theories based on fundamentals to theories based on bubbles.

As early as 1982, Dornbusch applied the notion of stochastic rational bubbles to the case of the strong dollar. According to this theory, there is a probability at any point in time that the bubble will burst during the subsequent period and the value of the currency will return to the equilibrium level determined by fundamentals. The differential in interest rates fully reflects and compensates for the possibility of the bubble bursting.

More recently it has been suggested that the dollar may in fact have been on an irrational bubble path. Two influential papers, written when the dollar was still near its peak--Marris (1985) and Krugman (1985)--argued that the mounting U.S. indebtedness to foreigners represented by record current account deficits would eventually force the dollar down sharply, and that this prospective depreciation was not correctly reflected in the small forward discount or interest differential (either short-term or long-term). "It appears that the market has simply not done its arithmetic, and has failed to realize that its expectations about continued dollar strength are not feasible" (Krugman (1985), p. 40).[6]

RATIONAL EXPECTATIONS AND THE FORWARD DISCOUNT

Meanwhile, evidence has continued to accumulate that the forward discount is a biased predictor of the future spot rate. A favorite way of explaining away such apparent statistical rejections of rational expectations is to appeal to the sort of "peso problem" that might arise in a speculative bubble. But, as explained in the following subsection, one of the present authors has presented calculations that tend to undermine the hypothesis that the dollar could have been on a single rational bubble from 1981 to 1985.[7] The expected probability of collapse that investors built in to the observed interest differential was high enough that it is very unlikely the dollar would have made it through four years without the bubble bursting, if that expectation was rational. This leaves the possibility of a bubble where the true probability of collapse may be different from the expected probability that investors build in to the forward discount.

Both Krugman and Marris have mentioned as partial support for their claim that the foreign exchange market may not be rational the large econometric literature that statistically rejects the hypothesis that the forward discount (or equivalently, by covered interest parity, the interest differential) is an unbiased predictor of the future spot rate. The most common test in this literature is a regression of the ex post change in the spot exchange rate against the forward discount at the beginning of the period. Under the null hypothesis the coefficient should be unity. But most authors have rejected the null hypothesis, finding that the coefficient is much closer to zero, and some even finding that the coefficient is of the incorrect sign. The implication is that one could expect to make money by betting against the forward discount whenever it is non-zero.[8] Bilson (1981) interprets this finding as "excessive speculation:" investors would do better if they would routinely reduce toward zero the magnitude of their expectations of exchange rate changes.

This forward market finding poses a puzzle in the context of the Krugman-Marris characterization of the dollar. It implies that as of 1985 (or for that matter at any time over the previous five years) the rationally expected rate of future dollar depreciation was less than the 3 percent a year implied in the forward discount.[9] The Krugman-Marris argument was that the rationally expected rate of future dollar depreciation would be much greater than the 3 percent a year implicit (against the mark or yen) in the market.[10] If we are to allow expectations to fail to be rational, we must somehow reconcile the two conflicting kinds of failure.

More discussion of the alleged bias in the forward exchange market is required. Most of the literature (for example the papers cited in footnote 8) does not interpret the finding as necessarily rejecting the hypothesis of rational expectations. Two other possible explanations are routinely offered: the existence of a risk premium and the "peso problem." We believe that, while both factors can be very important in other contexts, neither explains the systematic prediction errors made by the forward market during the strong-dollar period. We consider the risk premium briefly here, and the peso problem in the next subsection.

The first possible explanation is that the systematic component of the apparent prediction errors is really a risk premium separating the forward rate from investors' true expectations. It is a difficult argument either to refute or confirm, because expectations are not directly observable.

There are few sources of information to help isolate the risk premium out of the prediction errors made by the forward discount. One promising possibility is the surveys of market participants' exchange rate expectations conducted by the Economist's *Financial Report* and the *American Express Bank Review*.[11] The surveys allow us to measure expectations without the interference of the risk premium. In Frankel and Froot (1985) and Froot and Frankel (1986), we showed that those data for the 1981-85 period reflect a considerably greater expectation of dollar depreciation than do the forward discount or interest differential. (The biyearly averages are reported in rows 13-18 of Table 2.1.) We repeated standard tests of unbiasedness in expected depreciation and found even more significant rejections when the survey data, which must be free from any risk premium, are used than when the forward discount is used. First, we found unconditional bias: one would have persistently made money over the period June 1981-March 1985 by following the rule "buy and hold dollars." A related finding was that expectations were formed regressively--that is, the expected future spot rate puts some weight on a long-run equilibrium rate--but that the actual spot process did not bear out this expectation. Investors overestimated the speed of regression to a statistically significant degree.

An updating of the sample period to include data through December 1985 shows a dramatic shift in the nature of the bias: now it appears that investors on average underestimated the speed of regression toward long-run equilibrium to a statistically significant degree (Frankel and Froot (1987)). But the most robust finding, even with investors' expectations measured by the survey data instead of the forward discount, is excessive speculation in the sense of Bilson (1981): investors would have done better during the 1981-1985 period if they had routinely reduced their expectations of exchange rate changes. The rejection of rational expectations holds up even if one allows for measurement error in the survey data (provided it is random): one can reject the hypothesis that expectations are rational and that the apparent bias in the survey numbers is entirely attributable to measurement error. In addition, Froot and Frankel (1986) tests the hypothesis that no information about the risk premium is revealed in regressions of the ex post change in the spot rate on the forward discount. This hypothesis cannot be rejected, suggesting that the risk premium does not help explain why changes in the forward discount mispredict future changes in the spot rate. The rational expectations hypothesis appears in trouble.

AN EVALUATION OF THE SAFE-HAVEN AND RATIONAL BUBBLE HYPOTHESES

If the survey numbers are taken seriously as measuring investors' rate of expected depreciation, they imply a large *negative* risk premium paid on dollar assets during the 1981-85 period (a sharp decline from the near-zero risk premium in the 1970s). This is very different from the positive risk premium implied by standard tests of bias in the forward discount. Is a negative risk premium plausible nevertheless? Standard portfolio considerations would suggest not. The exchange risk premium in theory should depend on such variables as asset supplies and on return variances and covariances. The large U.S. government budget deficit and current account deficits mean that asset supplies should recently have been driving the dollar risk premium up, not down. One could posit an increase in the perceived riskiness of European currencies relative to the dollar, attributable for example to an increase in uncertainty regarding European monetary policy relative to U.S. monetary policy. But in that case it would be difficult to explain the increase in the U.S. interest differential after 1980; by itself a shift in demand toward U.S. assets due to uncertainty should have driven U.S. interest rates down.[12]

There is one explanation that has been seriously proposed for the dollar appreciation that is consistent with both a fall in the risk premium on dollars and an increase in the interest differential, in other words, consistent with the expected rate of depreciation increasing even more than the interest differential. That is the "safe haven" explanation mentioned above: an exogenous shift in demand toward U.S. assets due to perceptions of reduced country risk in the United States relative to abroad. According to this theory, risk has declined in the United States because of an improved business climate, in particular improved tax treatment for investment after 1981, which also explains the increase in U.S. real interest rates via an alleged investment boom.[13] Risk has increased in the rest of the world, not just because of debt problems in Latin America (which would alone not be relevant for the exchange rate or return differentials between the United States and Europe) but also because of political or country risk in Europe. Dooley and Isard (1985), for example, speak of a perceived threat of penalties on capital in Europe, "where the term 'penalty' is loosely defined to include formal taxation, the postponement of interest and principal payments, confiscation, destruction of property, and so forth."

We here propose a simple test be used to evaluate the safe haven hypothesis: a comparison of interest rates paid on securities that are physically located offshore, but that are denominated in dollars or otherwise covered on the forward exchange market to get around the problem of exchange risk, with interest rates paid on securities in the United States. That is, we are testing international closed, or covered, interest parity, not uncovered interest parity.

Tests of the offshore-onshore differential have been frequently employed to illustrate a number of points about the existence of capital controls or country risk: a negative differential for Germany until 1974 showed that capital controls discouraged capital inflow (Dooley and Isard (1980)); a positive differential for the United Kingdom until 1979 showed that capital controls discouraged outflow; positive differentials for France and Italy show that controls still discourage outflow (e.g.; Giavazzi and Pagano (1985), Classen and Wyplosz (1982)); a negative differential for Japan until 1979 showed that controls discouraged inflow (Otani and Tiwari (1981); Ito (1984) and Frankel (1984)); and, but for the foregoing exceptions, the generally small magnitude of differentials shows that capital mobility is very high among the major industrialized countries (e.g., Frenkel and Levich (1975), McCormick (1979), Boothe et al. (1985)).[14]

Table 2.2 reports mean daily differentials between offshore interest rates (covered) and domestic U.S. interest rates, for seven different pairs of securities. Remarkably, there was a relatively substantial positive differential in almost all cases, until recently, regardless whether one observes the offshore interest rate in the Euromarket, in the domestic U.K. market, or in the domestic German market.[15] From 1979 to 1982, the Euromarket rates exceeded the U.S. interbank rate by an average of about 100 basis points. A number of studies have noted that the Eurodollar rate does not move perfectly with the U.S. interbank or CD rate (Hartman (1983), Kreicher (1982)). They attribute the differential primarily to the fact that U.S. banks face reserve requirements against domestic deposits but not against Eurodeposits, so they are willing to pay a higher interest rate to depositors offshore. But the differential has been mostly swept under the rug in more general studies of covered interest parity.

Even those who have studied the Eurodollar-U.S. interbank differential treat it as a peculiarity of that particular market. This would make sense only if, on the one hand, the U.S. interbank rate were depressed below other U.S. interest rates (by U.S. reserve requirements) or if, on the other hand, Eurocurrency interest rates

TABLE 2.2
Deviations from Closed Interest Parity: Offshore Interest Rate (covered for exchange risk) Minus the United States Interest Rate (three-month interest rates in percentage per annum)

Offshore rate / U.S. rate	Euro-$ / T-Bill	Euro-$ / Interbank	Euro £ + fd / Interbank	U.K. ib + fd / Interbank	U.K. T-Bill + fd / T-Bill	Euro-DM + fd / Interbank	Ger. ib + fd / Interbank
Means							
Year							
1978	1.573	0.564	0.618	-0.840	-0.301	0.738	1.075
1979	1.894	0.786	0.886	0.622	1.656	1.047	1.491
1980	2.581	1.016	1.145	0.989	2.070	1.384	1.931
1981	2.190	0.923	1.080	1.085	2.105	1.242	1.778
1982	2.091	0.900	1.074	1.082	2.066	1.208	1.640
1983	0.660	0.546	0.676	0.691	0.577	0.786	1.127
1984	0.878	0.408	0.566	0.558	0.583	0.709	1.008
1985	0.571	0.295	0.414	0.410	0.305	0.396	0.622
Standard Deviations							
Year							
1978	0.666	0.262	0.390	0.846	0.975	0.477	0.484
1979	0.690	0.272	0.376	0.498	0.751	0.410	0.549
1980	1.027	0.371	0.785	0.795	1.233	0.526	0.565
1981	0.578	0.280	0.353	0.316	0.742	0.344	0.455
1982	0.736	0.205	0.242	0.223	0.746	0.308	0.357
1983	0.156	0.116	0.201	0.222	0.282	0.140	0.186
1984	0.401	0.078	0.143	0.134	0.418	0.194	0.234
1985	0.176	0.109	0.301	0.275	0.498	0.552	0.555

Note: ib ≡ interbank rate.
fd ≡ adjustment for the forward exchange discount.

were raised above domestic European interest rates (either by analogous reserve requirements in European countries or by perceived default risk in the Euromarket). But neither of these effects seems to hold. Table 2.2a shows small spreads between the Eurodollar rate and the Europound or Euromark rates (covered) or between them and the domestic U.K. and German interest rates. Indeed, Table 2.2 shows that the spread between covered pound or mark interest rates and domestic U.S. rates is even higher, and comes down even more after 1982, when Treasury bill rates are used than when banking rates are used. This finding contradicts the hypothesis that U.S. reserve requirements are the only factor driving a wedge between the Euromarket and the U.S. interbank market and that more direct arbitrage through other means works to reduce that wedge.

Why were foreigners and U.S. residents buying U.S. Treasury bills in 1979-1982 when they paid about 2 percent less than U.K. Treasury bills? The obvious response is that U.S. securities were preferred for safe-haven reasons. But since the differential predates the appreciation of the dollar, there is some difficulty in associating the two. This is particularly true after 1982, when the differential declines sharply. By 1985, when the dollar had appreciated much further, the Eurodollar rate was only 30 basis points above the domestic U.S. interbank interest rate, in the same range as the differentials for the pound, mark, yen, Canadian dollar, and Swiss franc. Chart 1 shows a comparison of the London Interbank Offer Rate (LIBOR) with a domestic U.S. CD rate, adjusted for reserve requirements. The differential, which was clearly positive in the early 1980s, peaked during the Mexican debt crisis in August 1982 and declined steadily afterward, reaching zero in early 1985, about the time when the dollar's value peaked. The evidence thus suggests that the United States was perceived as increasingly risky after 1982. The story based on safe-haven fundamentals does not explain the continued appreciation of the dollar from 1982 to February 1985 any better than the story based on real interest fundamentals. The field would appear to be open to bubble theories.

The possibility of speculative bubbles leads to the second explanation, besides the risk premium, that is often given for the econometric findings of biasedness in the forward exchange market: the peso problem. The standard tests presume that the error term, the difference between expected depreciation and the ex post realization, is distributed normally and independently over time. But if there is a small probability of a big decline in the value of the

Table 2.2a
Deviations from Interest Parity Within Jurisdictions
(Three-month interest rates in percentage per annum)

	Euro $ - fd / Euro	Euro £ / U.K. interbank	Euro £ / U.K. T-bill	Euro $ - fd / Euro DM	Euro DM / Ge. interbank
Means					
Year					
1978	-0.066	1.432	1.895	-0.187	-0.335
1979	-0.103	0.289	0.363	-0.220	-0.444
1980	-0.123	0.156	0.658	-0.373	-0.549
1981	-0.161	-0.004	0.228	-0.319	-0.525
1982	-0.179	0.003	0.207	-0.311	-0.431
1983	-0.131	-0.010	0.217	-0.239	-0.341
1984	-0.158	0.009	0.451	-0.300	-0.296
1985	-0.121	0.008	0.393	-0.100	-0.222
Standard Deviations					
Year					
1978	0.280	0.866	0.822	0.350	0.175
1979	0.272	0.288	0.466	0.408	0.253
1980	0.719	0.335	0.605	0.376	0.292
1981	0.286	0.250	0.470	0.250	0.317
1982	0.214	0.188	0.300	0.270	0.168
1983	0.179	0.143	0.240	0.088	0.113
1984	0.143	0.125	0.233	0.173	0.100
1985	0.285	0.119	0.418	0.552	0.094

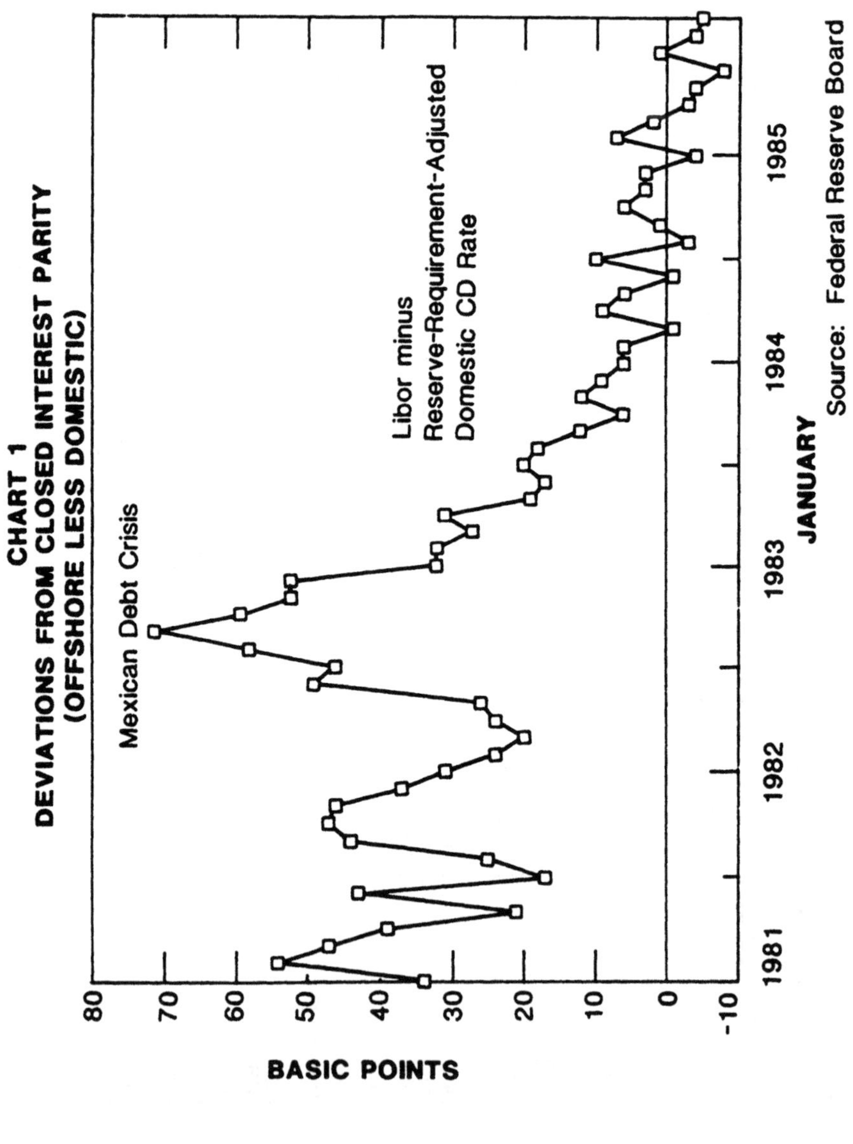
CHART 1
DEVIATIONS FROM CLOSED INTEREST PARITY
(OFFSHORE LESS DOMESTIC)
Mexican Debt Crisis
Libor minus
Reserve-Requirement-Adjusted
Domestic CD Rate
BASIC POINTS
80
70
60
50
40
30
20
10
0
-10
1981
1982
1983
1984
1985
JANUARY
Source: Federal Reserve Board

currency, the distributional assumption will not be met, the estimated standard errors will be incorrect, and unbiasedness may be spuriously rejected.[16] This problem is thought to be relevant for pegged currencies like the Mexican peso up until 1976, and generally less relevant for floating currencies. But if the dollar has been on a single speculative bubble path for four years, there could well be a small probability of a large decline in the form of a bursting of the bubble. It has been suggested that the forward discount may properly reflect that possibility, and that tests find a bias only because the event happens not to have occurred in the sample.

Calculations in Frankel (1985) tend to undermine the hypothesis that the forward discount during the period 1981-85 reflected rational expectations of a small probability of a large decline in the value of the dollar. Under the hypothesis that the bursting of the bubble would reverse half of the real appreciation of the dollar against the mark that has taken place since the 1970s, a 3 percent forward discount in March 1985 implied a 2.8 percent perceived probability of collapse during that month. One can multiply out the implied probabilities of non-collapse since January 1981, with no distributional assumptions needed, to find that the chance that such a bubble would have persisted for four years without bursting is only 3 percent. Thus the peso problem does not "get the forward exchange market off the hook." The period during which the forward discount was positive with no realized depreciation simply went on too long for the rational expectations hypothesis to emerge intact.

FUNDAMENTALISTS AND CHARTISTS

We can gather the conclusions reached so far into five propositions, each with elements of paradox.

(1) The dollar continued to rise even after all fundamentals (the interest differential, current account, etc.) apparently began moving the wrong way. The only explanation left would seem to be, almost tautologically, that investors were responding to a rising expected rate of change in the value of the dollar. In other words, the dollar was on a bubble path.

(2) Evidence suggests that the investor-expected rate of depreciation reflected in the forward discount is not equal to the rationally-expected rate of depreciation. The failure of a fall in the dollar to materialize in four years implies that the rationally-expected rate of depreciation was <u>less</u> than the forward discount.

(3) On the other hand, Krugman-Marris current account calculations suggested that the rationally-expected rate of depreciation was greater than the current forward discount.

(4) The survey data show that the respondents have since 1981 indeed held an expected rate of depreciation substantially greater than the forward discount. But interpreting their responses as true investor expectations, and interpreting the excess over the forward premium as a negative risk premium, raises a problem. If investors seriously expected the dollar to depreciate so fast, why did they buy dollars?

(5) In the safe-haven theory, a perceived shift in country risk rather than exchange risk might seem to explain many of the foregoing paradoxes. However, the covered differential between European and U.S. interest rates actually fell after 1982 suggesting that perceptions of country risk, if anything, shifted against the United States.

The model of fundamentalists and chartists that we are proposing has been designed to reconcile these conflicting conclusions. To begin with, we hypothesize that the views represented in the American Express and Economist 6-month surveys are primarily fundamentalist, like the views of Krugman and Marris (and most other economists). But it may be wrong to assume that investors' expected rate of depreciation is necessarily the one reported in the 6-month surveys or that there even is such a thing as "the" expected rate of depreciation (as most of our models do). Expectations are heterogeneous. Our model suggests that the market gives heavy weight to the chartists, whose expected rate of change in the value of the dollar has been on average much closer to zero, perhaps even positive. Paradox (4) is answered if fundamentalists' expectations are not the only ones determining positions that investors take in the market.

The increasing dollar overvaluation after the interest differential peaked in 1982 (measured short-term) or 1984 (measured long-term) would be explained by a falling market-expected rate of future depreciation (or rising expected rate of appreciation), with no necessary basis in fundamentals. The market-expected rate of depreciation declined over time, not necessarily because of any change in the expectations held by chartists or fundamentalists, but rather because of a shift in the weights assigned to the two by the portfolio managers. They are the agents who take positions in the market and determine the exchange rate. They gradually put less and less weight on the big-depreciation forecasts of the fundamentalists, as these forecasts continue to be proven false, and more and more weight on the chartists.

Before we proceed to show how such a model works, we offer evidence that there is not a single homogeneous expected rate of depreciation reflected in the survey data: the very short-term expectations (one-week and two-week) reported in a third survey of market participants, by Money Market Services, Inc., behave very differently from the medium-term expectations (3, 6, or 12 month) reported in any of the three surveys.[17]

EMPIRICAL RESULTS ON SHORT-TERM AND LONG-TERM EXPECTATIONS

One way of distinguishing empirically between the shorter- and longer-term expectations is to examine the weight survey respondents place on variables other than the contemporaneous spot rate in forming their expectations at different time horizons. Suppose, for example, that investors assign a weight of g to the lagged spot rate and a weight of 1-g to the current spot rate in forming their expectation of the future spot rate:

$$s^{m}_{t+1} = (1-g)s_t + gs_{t-1} \qquad (1)$$

where s_t is the logarithm of the current spot rate, and s_{t+1}^{m} is the market's expected future spot rate at time t. Subtracting s_t from both sides we have that expected depreciation is proportional to the current change in the spot rate:

$$\Delta s^{m}_{t+1} = -g\Delta s_{t+1} \qquad (2)$$

We term the model in equation (2) extrapolative expectations. If investors place positive weight on the lagged spot rate, so that g is positive, then equation (1) says that investors' expected future spot rate is a simple distributed lag. On the other hand, if investors tend to extrapolate the most recent change in the spot rate, so that g is negative, then equation (2) may be termed "bandwagon" expectations. We might, for instance, associate the fundamentalist viewpoint with a tendency to expect a currency which has recently appreciated to depreciate in the future (g > 0), and the chartist viewpoint with a tendency to expect on average some continuation of the past trend (g < 0).

Table 2.3 reports regression estimates of equation (2), using the survey expected depreciation as the lefthand-side variable.[18] The findings are ordered by the forecast horizon, from the shortest-term 1 and 2 week expectations, to the longer-term 12 month expectations. It is immediately evident that the shorter term expectations--1 week, 2 weeks and 1 month--all exhibit significant bandwagon tendencies: that $g < 0$. In the 1 week expectations, for example, an appreciation of 10 percent over the past week by itself generates the expectation that the spot rate will appreciate another 1.35 percent in the next seven days. This result is characteristic of destabilizing expectations, in which a current appreciation generates self-sustaining expectations of future appreciation.

In contrast with the shorter-term expectations, the longer-term results all point toward stabilizing distributed lag expectations. Each of the regressions at the 6 and 12 month forecast horizons estimates g to be significantly greater than zero.[19] The Economist 12 month data, for example, imply that a current 10 percent appreciation by itself generates an expectation of a 2.02 percent depreciation over the coming 12 months. Thus longer-term expectations feature a strongly positive weight on the lagged spot rate rather than complete weight on the contemporaneous spot rate, and in this sense they are stabilizing.

A second popular specification for the expected future spot rate is that it is a weighted average of the current spot rate and the (log) long-run equilibrium spot rate, $\bar{s}_t$:

$$s^m_{t+1} = (1-\theta)s_t + \theta\bar{s}_t \tag{3}$$

or in terms of expected depreciation:

$$\Delta s^m_{t+1} = \theta(\bar{s}_t - s_t) \tag{4}$$

If θ is positive, as, for example, in the Dornbusch overshooting model, the spot rate is expected to move in the direction of $\bar{s}_t$. Expectations are therefore regressive. This formulation for expectations is perhaps closest to the fundamentalists' view, because the long-run equilibrium to which investors expect the spot rate to return, $\bar{s}_t$, is determined by (fundamental) factors in the real economy. Alternatively, a finding of $\theta < 0$ implies that investors expect the spot rate to move away from the long-run equilibrium.

Table 2.4 presents tests of equation (4). Once again, there is strong evidence that shorter-term expectations are formed in a different manner than longer-term expectations. The shorter

TABLE 2.3
Extrapolative Expectations (Independent variable: $s_{t-1} - s_t$)

SUR Regressions (1) of Survey Expected Depreciation: $s^m_{t+1} - s_t = a+g(s_{t-1}-s_t)$

Data Set	Dates	coefficient $\hat{g}$	t:g=0	DW(2)	DF	R^2
MMS 1 Week	10/84-2/86	-0.1345 (0.0254)	-5.30***	1.89	239	0.76
MMS 2 Week	1/83-10/84	-0.0565 (0.0267)	-2.12**	1.76	179	0.33
MMS 1 Month	10/84-2/86	-0.0536 (0.0217)	-2.47**	1.48	171	0.40
Economist 3 Month	6/81-12/85	0.0416 (0.0210)	1.98*	1.81	184	0.30
MMS 3 Month	1/83-10/84	-0.0391 (0.0168)	-2.32**	1.49	179	0.37
Economist 6 Month	6/81-12/85	0.0730 (0.0225)	3.25***	1.36	184	0.54
Amex 6 Month	1/76-8/85	0.2994 (0.0487)	6.15***	1.89	45	0.81
Economist 12 Month	6/81-12/85	0.2018 (0.0296)	6.82***	1.47	184	0.84
Amex 12 Month	1/76-8/85	0.3796 (0.0798)	4.76***	0.94	45	0.72

(1) Amex 6 and 12 Month regressions use OLS due to the small number of degrees of freedom.
(2) The DW statistic is the average of the equation by equation OLS Durbin-Watson statistics for each data set. *Represents significance at the 10 percent level. **Represents significance at the 5 percent level. ***Represents significance at the 1 percent level. R^2 corresponds to an F test on all nonintercept parameters. Some of the above results are reported in Frankel and Froot (1987). Constant terms for each currency were included in the regressions, but not reported above.

TABLE 2.4
Regressive Expectations II (Independent variable: $\bar{s}_t - s_t$)
Long Run Equilibrium PPP

SUR Regressions (1) of Survey Expected Depreciation:

$$s^m_{t+1} - s_t = a + \theta(\bar{s}_t - s_t)$$

Data Set	Dates	coefficient $\hat{\theta}$	$t:\theta=0$	DW(2)	DF	R^2
MMS 1 Week	10/84-2/86	-0.0283 (0.0080)	-3.53***	2.10	219	0.58
MMS 2 Week	1/83-10/84	-0.0299 (0.0079)	-3.78***	2.15	179	0.61
MMS 1 Month	10/84-2/86	-0.0782 (0.0134)	-5.84***	1.40	151	0.79
Economist 3 Month	6/81-12/85	0.0223 (0.0126)	1.78*	1.66	184	0.26
MMS 3 Month	1/83-10/84	-0.0207 (0.0146)	-1.41	1.55	179	0.18
Economist 6 Month	6/81-12/85	0.0600 (0.0159)	3.77***	1.32	184	0.61
Amex 6 Month	1/76-8/85	0.0315 (0.0202)	1.56	1.22	45	0.21
Economist 12 Month	6/81-12/85	0.1750 (0.0216)	8.10***	1.25	184	0.88
Amex 12 Month	1/76-8/85	0.1236 (0.0276)	4.48***	0.60	45	0.69

(1) Amex 6 and 12 Month regressions use OLS due to the small number of degrees of freedom.
(2) The DW statistic is the average of the equation by equation OLS Durbin-Watson statistics for each data set. *Represents significance at the 10 percent level. **Represents significance at the 5 percent level. ***Represents significance at the 1 percent level. R^2 corresponds to an F test on all nonintercept parameters. Some of the above results are reported in Frankel and Froot (1987). Constant terms for each currency were included in the regressions, but not reported above.

forecast horizons all yield estimates of θ that are negative, additional evidence that shorter term speculation may be destabilizing. Indeed, the 1 week data suggest that the contemporaneous deviation from the long-run equilibrium is expected on average to grow by 3 percent over the subsequent seven days. In other words, short-term expectations are explosive. The significantly positive estimates of θ in the longer-term data sets suggest by contrast the longer-term expectations are strongly regressive. In the Economist 12 month data, for example, respondents expect any current deviation from the long-run equilibrium to decay by 17.5 percent over the following 12 months.

The final specification we consider is adaptive expectations. In this case, agents are hypothesized to form their expectation of the future spot rate as a weighted average of the current spot rate and the lagged expected spot rate:

$$s^m_{t+1} = (1-\gamma)s_t + \gamma s^m_t \tag{5}$$

Expected depreciation is now proportional to the contemporaneous prediction error:

$$\Delta s^m_{t+1} = (s^m_t - s_t) \tag{6}$$

Table 2.5 reports estimates of equation (6). The R^2 statistics are generally lower than in Tables 2.3 and 2.4, suggesting that the surveys are not characterized as well by adaptive expectations as they are by regressive and extrapolative expectations. Nevertheless, the results are qualitatively comparable with those of the previous two tables. The shorter-term expectations place significantly negative weight on the lagged expectation. At the same time there is evidence that the longer-term data place positive weight on the lagged expectation, that longer-term expectations are adaptive.

The results of Tables 2.3, 2.4 and 2.5 suggest that in all three of our standard models of expectations--extrapolative, regressive and adaptive--short-term and long-term expectations behave very differently from one another. In terms of the distinction between fundamentalists and chartists views, we associate the longer-term expectations, which are consistently stabilizing, with the fundamentalists, and the shorter-term forecasts, which seem to have a destabilizing nature, with the chartist expectations. Within each of the above tables, it is as if there are actually two models of expectations operating, one at each end of the spectrum of forecast horizons, and a blend in between. Under this view, respondents use

TABLE 2.5
Adaptive Expectations (Independent variable: $s_t^m - s_t$)

SUR Regressions(1) of Survey Expected Depreciation:
$s_{t+1}^m - s_t = a + \gamma(s_t^m - s_t)$

Data Set	Dates	coefficient γ	t:γ=0	DW(2)	DF	R^2
MMS 1 Week	10/84-2/86	-0.1047 (0.0256)	-4.09***	1.69	211	0.65
MMS 2 Week	1/83-10/84	-0.0296 (0.0255)	-1.16	1.68	175	0.13
MMS 1 Month	10/84-2/86	0.0121 (0.0235)	0.52	1.31	135	0.03
Economist 3 Month	6/81-12/85	0.0798 (0.0203)	3.93***	2.01	169	0.63
MMS 3 Month	1/83-10/84	-0.0272 (0.0215)	-1.27	1.29	159	0.15
Economist 6 Month	6/81-12/85	0.0516 (0.0161)	3.20***	1.12	159	0.53
Amex 6 Month	1/76-8/85	-0.0702 (0.1200)	-0.59	2.10	15	0.04
Economist 12 Month	6/81-12/85	-0.0093 (0.0244)	-0.38	1.10	139	0.02
Amex 12 Month	1/76-8/85	0.0946 (0.0212)	4.46***	0.55	31	0.69

(1) Amex 6 and 12 Month regressions use OLS due to the small number of degrees of freedom.
(2) The DW statistic is the average of the equation by equation OLS Durbin-Watson statistics for each data set. *Represents significance at the 10 percent level. **Represents significance at the 5 percent level. ***Represents significance at the 1 percent level. R^2 corresponds to an F test on all nonintercept parameters. Some of the above results are reported in Frankel and Froot (1987). Constant terms for each currency were included in the regressions, but not reported above.

some weighted average of the chartist and fundamentalist forecasts in formulating their expectations for the value of the dollar at a given future date, with the weights depending on how far off that date is.

These results suggest an alternative interpretation of how chartist and fundamentalist views are aggregated in the marketplace, an aggregation that takes place without the benefit of portfolio managers. It is possible that the chartists are simply people who tend to think short-term and the fundamentalists are people who tend to think long-term. For example, the former may by profession be "traders", people who buy and sell foreign exchange on a short-term basis and have evolved different ways of thinking than the latter, who may by profession buy and hold longer-term securities.[20]

In any case, one could interpret the two groups as taking positions in the market directly, rather than merely issuing forecasts for the portfolio managers to read. The market price of foreign exchange would then be determined by demand coming from both groups. But the weights that the market gives to the two change over time, according to the groups' respective wealths.[21] If the fundamentalists sell the dollar short and keep losing money, while the chartists go long and keep gaining, in the long run the fundamentalists will go bankrupt and there will only be chartists in the marketplace. The model that we develop in the next section pursues the portfolio manager's decision-making problem instead of the marketplace-aggregation idea, but the two are similar in spirit.

Yet another possible interpretation of the survey data is that the two ways of thinking represent conflicting forces within the mind of a single representative agent. When respondents answer the longer-term surveys they give the views that their economic reason tells them are correct. When they get into the trading room they give greater weight to their instincts, especially if past bets based on their economic reason have been followed by ruinous "negative reinforcement." A respondent may think that when the dollar begins its plunge, he or she will be able to get out before everyone else does. This opposing instinctual force comes out in the survey only when the question pertains to the very short-term--one or two weeks; it would be too big a contradiction for his conscience if a respondent were to report a one-week expectation of dollar depreciation that was (proportionately) just as big as the answer to the 6-month question, at the same time that he or she was taking a long position in dollars. Again, we prefer the interpretation where the survey reflects the true expectations of the respondent, and the market trading is done by some higher authority; but others may prefer the more complex psychological interpretation.

The fragments of empirical evidence in Tables 2.3, 2.4, and 2.5 are the only ones we will offer by way of testing our approach. The aim in what follows is to construct a model that reconciles the apparent contradictions discussed in Part 1. There will be no further hypothesis testing.

AN ESTIMATE OF THE WEIGHTS

We think of the value of the dollar as being driven by the decisions of portfolio managers who use a weighted average of the expectations of fundamentalists and chartists. Specifically,

$$\Delta s^{m}_{t+1} = \omega_t \Delta s^{f}_{t+1} + (1-\omega_t)\Delta s^{c}_{t+1} \tag{7}$$

where Δs_{t+1}^{m} is the rate of change in the spot rate expected by the portfolio managers, Δs_{t+1}^{f} and Δs_{t+1}^{c} are defined similarly for the fundamentalists and chartists, and ω_t is the weight given to fundamentalist views. For simplicity we assume $\Delta s_{t+1}^{c} = 0$. Thus equation (7) becomes

$$\Delta s^{m}_{t+1} = \omega_t \Delta s^{f}_{t+1} \tag{8}$$

or

$$\omega_t = \frac{\Delta s^{m}_{t+1}}{\Delta s^{f}_{t+1}}$$

If we take the 6-month forward discount to be representative of portfolio managers' expectations and the 6-month survey to be representative of fundamentalists' expectations, we can get a rough idea of how the weight, ω_t, varies over time.

Table 2.6 contains estimates of ω_t from the late 1970s to 1985. (There are, unfortunately, no survey data for 1980.) The table indicates a preponderance of fundamentalism in the late seventies; portfolio managers gave almost complete weight to this view. But beginning in 1981, as the dollar began to rise, the forward discount increased less rapidly than fundamentalists' expected depreciation, indicating that the market (the portfolio managers in our story) was beginning to pay less attention to the fundamentalists' view. By 1985, the market's expected depreciation had fallen to about

TABLE 2.6
Estimated Weights Given to Fundamentalists by Portfolio Managers

	Year					
	1976–79	1981	1982	1983	1984	1985
Forward Discount fd	1.06	3.74	3.01	1.10	3.07	-0.16
Survey Expected Depreciation Δs^m_{t+1}	1.20	8.90	10.31	10.42	11.66	4.00
$\omega_t \equiv (fd/\Delta s^m_{t+1})$	0.88	0.42	0.29	0.11	0.26	-0.04

Notes: Forward discount, 1976-85 is at 6 months and includes data through September 1985 for the average of five currencies, the pound, French franc, mark, Swiss franc and yen. Survey expected depreciation 1981-85 is from the Economist 6 month survey data, and for 1976-79 is from the AMEX survey data for the same five currencies.

zero. According to these computations, fundamentalists were being completely ignored.

While the above scenario solves the paradox posed in proposition (4), it leaves unanswered the question of how the weight $_t$, which appears to have fallen dramatically since the late 1970s, is determined by portfolio managers. Furthermore, if portfolio managers have small risk premia, and thus expect depreciation at a rate close to that predicted by the forward discount, we still must account for the spectacular rise of the dollar (proposition (1)), and resolve how the rationally expected depreciation differs from the forward discount (propositions (2) and (3)).

PORTFOLIO MANAGERS AND EXCHANGE RATE DYNAMICS

Up to this point we have characterized the chartist and fundamentalist views of the world, and hinted at the approximate mix that portfolio managers would need to use if the market risk premium is to be near zero. We now turn to an examination of the behavior of portfolio managers, and to the determination of the equilibrium spot rate. In particular, we first focus exclusively on the dynamics of the spot rate which are generated by the changing expectations of portfolio managers. We then extend the framework to include the evolution of fundamentals which eventually must bring the dollar back down.

DETERMINATION OF THE EXCHANGE RATE

A general model of exchange rate determination can be written:

$$s_t = c\Delta s^m_{t+1} + z_t \tag{9}$$

where s(t) is the log of the spot rate, Δs^m_{t+1} is the rate of depreciation expected by "the market" (portfolio mangers) and z_t represents other contemporaneous determinants. This very general formulation, in which the first term can be thought of as speculative factors and the second as fundamentals, has been used by Mussa (1976) and Kohlhagen (1979). An easy way to interpret equation (9) is in terms of the monetary model of Mussa (1976), Frenkel (1976) and Bilson (1978). Then c would be interpreted as the semi-elasticity of money demand with respect to the alternative rate of return (which could be the interest differential, expected depreciation or expected inflation differential; as noted in section 1.1, the three are equal if uncovered interest parity and purchasing power parity hold), and z_t would be interpreted as the log of the domestic money supply relative to the foreign (minus the log of relative income, or any other determinants of real money demand). An interpretation of equation (9) in terms of the portfolio-balance approach is slightly more awkward because of nonlinearity. But we could define:

$$z_t = d_t - f_t - c(i_t - i^*_t) \tag{10}$$

where d_t is the log of the supply of domestic assets including not only money but also bonds and other assets, f_t is the log of the supply of foreign assets, and $i_t - i^*_t$ is the nominal interest differential. Then equation (9) can be derived as a linear approximation to the solution for the spot rate in a system where the share of the portfolio allocated to foreign assets depends on the expected return differential or risk premium, $i_t - i^*_t - \Delta s^m_{t+1}$. If investors diversify their portfolios optimally, c can be seen to depend inversely on the variance of the exchange rate and the coefficient of relative risk-aversion.[22] In any case, the key point behind equation (9), common throughout the asset-market view of exchange rates, is that an increase in the expected rate of future depreciation will reduce demand for the currency today, and therefore will cause it to depreciate today.

The present paper imbeds in the otherwise standard asset pricing model given by equation (9) a form of market expectations that follows equation (7). That is, we assume that portfolio managers' expectations are a weighted average of the expectations of fundamentalists, who think the spot rate regresses to long-run equilibrium, and the expectations of chartists who use time series methods:

$$\Delta s^m_{t+1} = \omega_t \Delta s^f_{t+1} + (1-\omega_t)\Delta s^c_{t+1} \tag{11}$$

We define $\bar{s}$ to be the logarithm of the long-run equilibrium rate and θ to be the speed of regression of s_t to $\bar{s}$. In the view of fundamentalists:

$$\Delta s^f_{t+1} = \theta(\bar{s} - s_t) \tag{12}$$

In the context of some standard versions of equation (9)--the monetary model of Dornbusch (1976) in which goods prices adjust slowly over time or the portfolio-balance models in which the stock of foreign assets adjusts slowly over time--it can be shown that equation (12) might be precisely the rational form for expectations to take if there were no chartists in the market, $\omega_t = 1$. Unfortunately for the fundamentalists, the distinction is crucial; equation (12) will not be rational given the complete model.

For example, if we define z_t in equation (9) as the interest differential we have:

$$s_t = a + c\theta(\bar{s} - s_t) - b(i_t - i_t^*) \qquad (13)$$

Uncovered interest parity, $i_t - i_t^* = \theta(\bar{s} - s(t))$, implies that $\theta = 1/(\beta - c)$ and $a = \bar{s}$. It is then straightforward to show that θ can be rational within the Dornbusch (1976) overshooting model.[23]

In the second group of models (Kouri (1976) and Rodriguez (1980) are references), overshooting occurs because the stock of net foreign assets adjusts slowly through current account surpluses or deficits. A monetary expansion creates an imbalance in investors' portfolios which can be resolved only by an initial increase in the value of net foreign assets. This sudden depreciation of the domestic currency sets in motion an adjustment process in which the level of net foreign assets increases and the currency appreciates to its new steady-state level. In such a model (which is similar to the simulation model below), the rate of adjustment of the spot rate, θ, may also be rational, if there are no chartists. Repeating equation (13) but using the log of the stock of net foreign assets instead of the interest differential as the important fundamental, we have in continuous time:

$$s(t) = a + c\theta(\bar{s} - s(t)) - df(t) \qquad (14)$$

Suppose the actual rate of depreciation is $s(t) = \nu(\bar{s} - s(t))$. Equation (14) then can be rewritten in terms of deviations from the steady-state levels of the exchange rate and net foreign assets, $\bar{s}$ and $\bar{f}$.

$$\dot{s}(t) = \frac{-\nu}{c\theta}(\bar{s} - s(t)) - \frac{d\nu}{c\theta}(\bar{f} - f(t)) \qquad (15)$$

where rationality implies that $\nu = \theta$. Following Rodriguez (1980), the normalized current account surplus may also be expressed in deviations from steady-state equilibrium:

$$\dot{f} = -q(\bar{s} - s(t)) + \gamma(\bar{f} - f(t)) \tag{16}$$

where q and γ are the elasticities of the current account with respect to the exchange rate and the level of net foreign assets, respectively. The system of equations (15) and (16) then has the rational expectations solution:

$$\theta = \frac{c\gamma - 1 + [1-c\gamma)^2 + 4c(\gamma + dq)]^{\frac{1}{2}}}{2c} \tag{17}$$

THE MODEL WITH EXOGENOUS FUNDAMENTALS

We now turn to describe the complete model, assuming for the time being that important fundamentals remain fixed. Regardless of which specification we use for the fundamentals, the existence of chartists whose views are given time-varying weights by the portfolio managers complicates the model. For simplicity, we study the case in which the chartists believe the exchange rate follows a random walk, $\Delta s^c_{t+1} = 0$. Thus equation (7) becomes:

$$\Delta s^m_{t+1} = \omega_t \theta(\bar{s} - s_t) \tag{17a}$$

Since the changing weights by themselves generate self-sustaining dynamics, the expectations of fundamentalists will no longer be rational, except for the trivial case in which fundamentalist and chartist expectations are the same, $\theta = 0$.

The "bubble" path of the exchange rate will be driven by the dynamics of portfolio managers' expected depreciation. We assume that the weight given to fundamentalist views by portfolio managers, ω_t, evolves according to:

$$\Delta\omega_t = \delta(\hat{\omega}_{t-1} - \omega_{t-1}) \tag{18}$$

$\hat{\omega}_{t-1}$ is in turn defined as the weight, computed _ex post_, that would have accurately predicted the contemporaneous change in the spot rate, defined by the equation:

$$\Delta s_t = \hat{\omega}_{t-1}\theta(\bar{s}-s_{t-1}) \tag{19}$$

Equations (18) and (19) give us:

$$\Delta\omega_t = \delta\frac{\Delta s_t}{\theta(\bar{s}-s_{t-1})} - \delta\omega_{t-1} \tag{20}$$

The coefficient δ in equation (20) controls the adaptiveness of ω_t.

One interpretation for δ is that it is chosen by portfolio managers who use the principles of Bayesian inference to combine prior information with actual realizations of the spot process. This leads to an expression for δ which changes over time. To simplify the following analysis we assume that δ is constant; in the first appendix we explore more precisely the problem that portfolio managers face in choosing δ. The results that emerge there are qualitatively similar to those that follow here.

Taking the limit to continuous time, we can rewrite equation (20) as:

$$\dot{\omega}(t) = \delta\left[\frac{\dot{s}(t)}{\theta(\bar{s}-s(t))} - \omega(t)\right] \quad \text{if } 0<\omega(t)<1 \tag{21}$$

$$\text{if } \omega(t)=0 \text{ then } \begin{cases} \dot{\omega}(t)=0 & \text{if } \dot{s}(t)\leq 0 \\ \dot{\omega}(t)=\dfrac{\delta\dot{s}(t)}{\theta(\bar{s}-s)} & \text{if } \dot{s}(t)>0 \end{cases} \tag{21a}$$

$$\text{if } \omega(t)=1 \text{ then } \begin{cases} \dot{\omega}(t)=0 & \text{if } \dot{s}(t)\geq\theta(\bar{s}-s(t)) \\ \dot{\omega}(t)=\dfrac{\delta\dot{s}(t)}{\theta(\bar{s}-s(t))}-\delta & \text{if } \dot{s}(t)<\theta(\bar{s}-s(t)) \end{cases} \tag{21b}$$

where a dot over a variable indicates the total derivative with respect to time. The restrictions that are imposed when $\omega(t) = 0$ and $\omega(t) = 1$ are to keep $\omega(t)$ from moving outside the interval [0,1]. These restrictions are in the spirit of the portfolio managers choice

FIGURE 1

SIMULATED VALUE OF THE DOLLAR ABOVE ITS LONG RUN EQUILIBRIUM

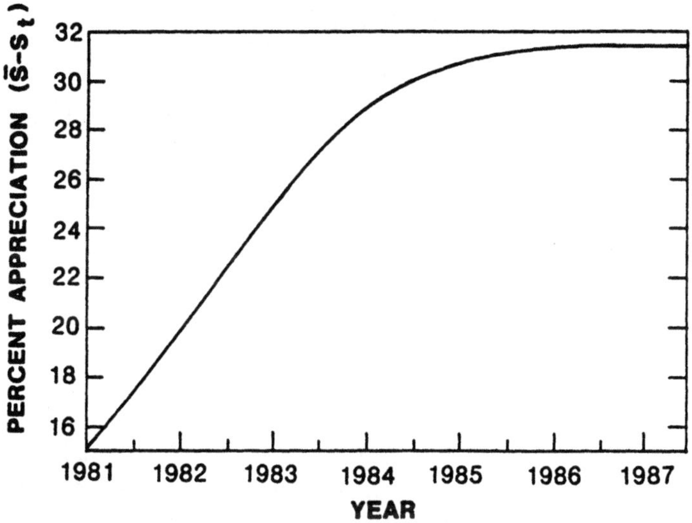

FIGURE 2

SIMULATED WEIGHT PLACED ON FUNDAMENTALIST EXPECTATIONS BY PORTFOLIO MANAGERS

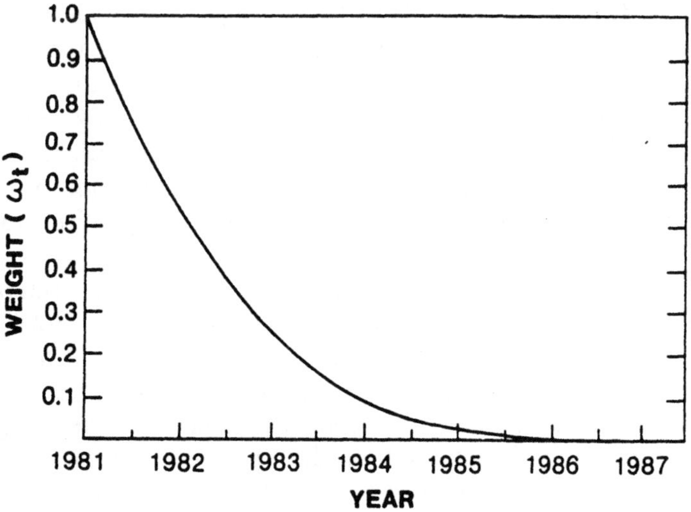

set: the portfolio manager can at most take one view or the other exclusively.

The evolution of the spot rate can be expressed by taking the derivative of equation (9) (for now holding z and the long-run equilibrium, $\bar{s}$, constant):

$$\dot{s}(t) = \left[\frac{\dot{\omega}(t)c\theta}{1 + c\theta\omega(t)}\right](\bar{s} - s(t)) \tag{22}$$

Equations (21) and (22) can be solved simultaneously and rewritten, for interior values of ω, as:

$$\dot{\omega}(t) = \frac{-\delta\omega(t)\,(1 + c\theta\,\omega(t))}{1 + c\theta\omega(t) - \delta c} \quad \text{if } 0 < \omega(t) < 1 \tag{23}$$

$$\dot{s}(t) = \left[\frac{-\delta\omega(t)c\vartheta}{1 + c\theta\omega(t) - \delta c}\right](\bar{s} - s(t)) \tag{24}$$

In principle, an analytic solution to the differential equation (23) could be substituted into (24), and then (24) could be integrated directly.[24] For our purposes it is more desirable to use a finite difference method to simulate the motion of the system. In doing so we must pick values for the coefficients, c, θ and δ, and starting values for ω(t) and s(t).

To exclude any unreasonable time paths implied by equations (23) and (24), we impose the obvious sign restrictions on the coefficients. The parameter θ must be positive and less than one if expectations are to be regressive, that is, if they are to predict a return to the long-run equilibrium at a finite rate. By definition, δ and ω(t) lie in the interval [0,1] since they are weights. The coefficient c measures the responsiveness of the spot rate to changes in expected depreciation and must be positive to be sensible.

These restrictions, however, are not enough to determine unambiguously the sign of the denominator of equations (23) and (24). The three possibilities are that: $1 + c\,\theta\omega(t) - \delta c < 0$ for all ω; $1 \lesseqgtr 0$ as $\omega(t) \lesseqgtr \omega^*$, where $0 < \omega^* < 1$. If $1 + c\,\theta\omega(t) - \delta c < 0$, the system will be stable and will tend to return to the long-run equilibrium from any initial level of the spot rate. This might be the case if portfolio managers use only the most recent realization of the spot rate to choose ω(t), that is, if $\delta \approx 1$. If, on the other hand, portfolio managers give substantial weight to prior information so that δ is small, the expression $1 + c\,\theta\omega(t) - \delta c$ will be positive.

will be positive. In this case the spot rate will tend to move away from the long-run equilibrium if it is perturbed.[25]

Let us assume that portfolio managers are slow learners.[26] What does this assumption imply about the path of the dollar? If we take as a starting point the late 1970s, when $s(t) \approx \bar{s}$ and when $\omega_t \approx 1$ (as the calculations presented in Table 2.6 suggest), equation (24) says that the spot rate is in equilibrium, that $\dot{s}(t) = 0$. From equation (21b), we see that $\dot{\omega}(t) = 0$ as well. Thus the system is in a steady-state equilibrium, with market expectations exclusively reflecting the views of fundamentalists.

But given that $1 + c\,\theta\omega(t) - \delta c > 0$, this equilibrium is unstable, and any shock starts things in motion. Suppose that there is an unanticipated appreciation (the unexpected persistence of high long-term US interest rates in the early 1980s, for example). The sign restrictions imply that $\omega(t)$ is unambiguously falling over time. Equation (23) says that the chartists are gaining prominence, since $\dot{\omega}(t) < 0$. The exchange rate begins to trace out a bubble path, moving away from long-run equilibrium; equation (24) shows that $\dot{s}(t) < 0$ when $\bar{s} > s(t)$. This process cannot, however, go on forever, because market expectations are eventually determined only by chartist views. At this point the bubble dynamics die out since both $\omega(t)$ and $\dot{\omega}(t)$ fall to zero. From equation (24), the spot rate then stops moving away from long-run equilibrium, as it approaches a new, higher equilibrium level where $\dot{s}(t) = 0$. In the words of Dornbusch (1983), the exchange rate is both high and stuck.

Figures 1 and 2 trace out a "base-case" simulation of the time profile of the spot rate and ω. They are intended only to suggest that the model can potentially account for a large and sustained dollar appreciation. The figures assume that the dollar is perturbed out of a steady state equilibrium where $\bar{s} = s(t)$ and $\omega(0) = 1$ in October 1980. The dollar rises at a decreasing rate until sometime in 1985, when, as can be seen in Figure 2, the simulated weight placed on fundamentalist expectations becomes negligible. A steady state obtains at a new higher level, about 31 percent above the long-run equilibrium implied by purchasing power parity. Although we tried to choose reasonable values for the parameters used in this example, the precise level of the plateau and the rate at which the currency approaches it are sensitive to different choices of parameters. In a second appendix, available on request, we give more detail on values used in the simulation.

It is worth emphasizing that the demand for dollars increases and the currency appreciates along its bubble path even though <u>none</u> of the actors expects appreciation. This result is due to the implicit stock adjustment taking place. As portfolio managers reject their fundamentalist roots, they reshuffle their portfolios to hold a greater share in dollar assets. For fixed relative asset supplies, a greater dollar share can be obtained in equilibrium only by additional appreciation. This unexpected appreciation, in turn, further convinces portfolio managers to embrace chartism. The rising dollar becomes self-sustaining. In the end when the spiral finally levels off at $\omega(t) = 0$, the level at which the currency becomes stuck represents a fully rational equilibrium: portfolio managers expect zero depreciation and the rate of change of the exchange rate is indeed zero.

The sense in which the model violates rational expectations can be seen by inspecting equation (24). Recall that market-expected depreciation, that of portfolio managers, is a weighed average of chartist and fundamentalist expectations, $\omega(t)\ \theta(\bar{s} - s(t))$. But the actual, or rational, expected rate of depreciation is given by $\left[\frac{-\delta c}{1 + c\theta\omega(t) - \delta c}\right]\omega(t)\theta(\bar{s} - s(t))$. The two are not equal, unless $\omega = 0$.[27] The problem we gave portfolio managers was to pick $\omega(t)$ in a way that best describes the spot process they observe (given the prior confidence they had in fundamentalist predictions). But theirs is a thankless task, since the spot process is more complicated.

THE MODEL WITH ENDOGENOUS FUNDAMENTALS

The results so far offer an explanation for the paradox of proposition (1), that sustained dollar appreciation occurs even though all agents expect depreciation. But a spot rate that is stuck at a disequilibrium level is an unlikely end for any reasonable story. The next step is to specify the mechanism by which the unsustainability of the dollar is manifest in the model.

The most obvious fundamental which must eventually force the dollar down is the stock of net foreign assets. Reductions in this stock, through large current account deficits, cannot take place indefinitely. Sustained borrowing would, in the long run, raise the level of debt above the present discounted value of income. But long before this point of insolvency is reached, the gains from a U.S. policy aimed at reducing the outstanding liabilities (either through direct taxes or penalties on capital, or through monetization) would increase in comparison to the costs. If foreigners associate large current account deficits with the potential for moral hazard, they would treat U.S. securities as increasingly risky and would force a decline in the level of the dollar.

To incorporate the effects of current account imbalances, we consider the model, similar to Rodriguez (1980), given in equation (14):

$$s_t = a + c\Delta s^m_{t+1} - df \tag{25}$$

where Δs^m_{t+1} is defined in equation (7a) and where f represents the log of cumulated US current account balances. The coefficient, d, is the semi-elasticity of the spot rate with respect to transfers of wealth, and must be positive to be sensible. The differential equations (23) and (24) now become:

$$\dot{\omega}(t) = \left[\frac{\delta}{1+c\theta\omega(t)-\delta c}\right]\left[-\omega(t)(1+c\theta\omega(t)) - \frac{d\dot{f}}{\theta(\bar{s}-s(t))}\right] \text{ if } 0<\omega(t)<1 \tag{26}$$

$$\dot{s}(t) = \frac{-\delta\omega(t)c\theta(\bar{s} - s(t)) + d\dot{f}}{1 + c\omega(t)\theta - \delta c} \tag{27}$$

If we were to follow the route of trying to solve analytically the system of differential equations, we would add a third equation giving the "normalized" current account, $\dot{f}$, as a function of s(t). (See, for example, equation (16) above.) But we here instead pursue the simulation approach.

In the simulation we use actual current account data for $\dot{f}$, the change in the stock of net foreign assets. Figures 3 and 4 trace out paths for the differential equations (26) and (27). During the initial phases of the dollar appreciation, the current account, which is thought to respond to the appreciation with a lag, does not noticeably affect the rise of the dollar. But as ω becomes small, the spot rate becomes more sensitive to changes in the level of the current account, and the external deficits of 1983-1985 quickly turn the trend. When ω is small and portfolio managers observe an incipient depreciation of the dollar, they begin to place more weight on the forecasts of fundamentalists, thus accelerating the depreciation initiated by the current account deficits. There is a "fundamentalist revival." Ironically, fundamentalists are initially driven out of the market as the dollar appreciates, *even though they are ultimately right about its return to s.*

Naturally, all of our results are sensitive to the precise parameters chosen. To gain an idea of the various sensitivities, we

FIGURE 3

SIMULATED VALUE OF THE DOLLAR ABOVE ITS LONG RUN EQUILIBRIUM

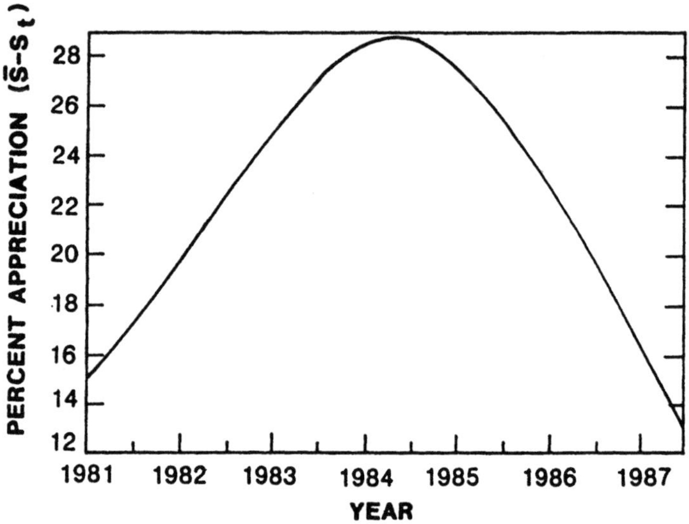

FIGURE 4

SIMULATED WEIGHT PLACED ON FUNDAMENTALIST EXPECTATIONS BY PORTFOLIO MANAGERS

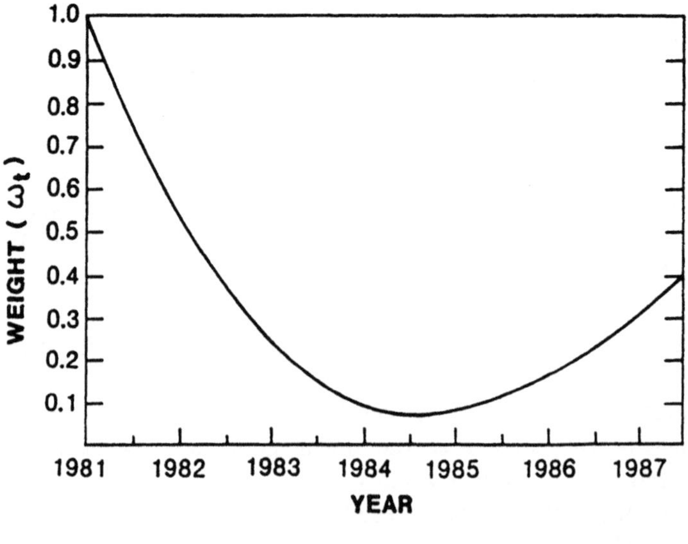

report in Table 2.7 results using alternative sets of parameter values in the simulation of Figure 3 (or equation (27)). While there is some variation, the qualitative pattern of bubble appreciation, followed by a slow turnaround and bubble depreciation, remains evident in all cases.

Recall that one of the main aims of the model is to account for the two seemingly contradictory facts given by propositions (2) and (3): first that market efficiency test results imply that the rationally expected rate of dollar depreciation has been less than the forward discount, and second that the calculations based on fundamentals, such as those by Krugman and Marris, imply that the rationally expected rate of depreciation, by 1985, became greater than the forward discount.

Table 2.8 clarifies how the model resolves this paradox. The first two lines show the expectations of our two forecasters, the chartists and fundamentalists. The third line repeats the six-month survey expectations to demonstrate that they may in fact be fairly well described by the simple regressive formulation we use to represent fundamentalist expectations in line two. The fourth line contains the expected depreciation of the portfolio managers. Note that these expectations are close to the forward discount in line six, even though the forecasts of the fundamentalists and of the chartists are not. Since only the portfolio managers are hypothesized to take positions in the market, we can say that the magnitude of the market risk premium is small (as mean-variance optimization would predict). Finally, line five shows the actual depreciation in the simulation, which is equivalent to the rationally expected depreciation given the model above. (Of course, none of the agents has the entire model in his information set.) Notice that during the 1981-1984 period, the rationally expected depreciation is not only significantly less than the forward discount, but less than zero. This pattern agrees with the results of market efficiency tests discussed earlier. But the rationally expected depreciation is increasing over time. Sometime in late 1984 or early 1985, the rationally expected rate of depreciation becomes positive and crosses the forward discount. As calculations of the Krugman-Marris type would indicate, rationally expected depreciation is ow greater than the forward discount. The paradox of propositions (2) and (3) is thus resolved within the model.

All this comes at what might seem a high cost: portfolio managers behave irrationally in that they do not use the entire model in formulating their exchange rate forecasts. But another interpretation of this behavior is possible, in that portfolio managers

TABLE 2.7
Sensitivity Analysis for the Simulation of the Dollar

Parameter				Maximum appreciation of the dollar above initial shock	Number of months until peak
delta	c	theta	d	(in percent)	
0.04	25	0.045	-0.005	11.4	41
0.06	25	0.045	-0.005	26.9	27
0.02	25	0.045	-0.005	5.8	44
0.04	15	0.045	-0.005	6.4	38
0.04	35	0.045	-0.005	18.1	40
0.04	25	0.03	-0.005	8.8	36
0.04	25	0.06	-0.005	13.5	44
0.04	25	0.045	0	16.4	80
0.04	25	0.045	-0.0025	11.6	45
0.04	25	0.045	-0.0075	11.4	38

Notes: These estimates correspond to the simulation depicted in figure 8. The parameter delta falls over time according to equation (19).

TABLE 2.8
Alternative Measures of Expected Depreciation (in percent per annum)

Expectation from:	Line	1981	1982	Year 1983	1984	1985	1986
Chartists in the simulation	(1)	0	0	0	0	0	0
Fundamentalists in the simulation	(2)	7.63	9.82	11.68	11.98	10.33	7.69
Economist 6 Month Survey	(3)	8.90	10.31	10.42	11.66	4.00	NA
Weighted Average Expected Depreciation in the Simulation	(4)	5.29	3.31	1.59	0.99	1.49	2.08
Rationally Expected Depreciation in the Simulation	(5)	-2.97	-5.61	-4.38	-0.72	3.89	6.22
Actual Forward Discount	(6)	3.74	3.01	1.10	3.07	-0.16	NA

Notes: Fundamentalists in the simulation use regressivity parameter of .045, implying that about 70% of the contemporaneous overvaluation is expected to remain after one year. The Economist 6 month survey includes data through April 1985. Weighted average expected depreciation in the simulation is a weighted average of chartists and fundamentalists, where the weights are those of portfolio managers. Rationally Expected Depreciation is the perfect foresight solution given by equations (19) and (20). The actual 6 month forward discount includes data through September 1985.

are actually doing the best they can in a confusing world. Within this framework they cannot have been more rational; abandoning fundamentalism more quickly would not solve the problem in the sense that their expectations would not be validated by the resulting spot process in the long run. In trying to learn about the world after a regime change, our portfolio managers use convex combinations of models which are already available to them and which have worked in the past. In this context, rationality is the rather strong presumption that one of the prior models is correct. It is hard to imagine how agents, after a regime change, would know the correct model.

CONCLUSIONS AND EXTENSIONS

This paper has posed an unorthodox explanation for the recent acrobatics of the dollar. The model we use assumes less than fully rational behavior in the sense that none of the three classes of actors (chartists, fundamentalists and portfolio managers) conditions its forecasts on the full information set of the model. In effect, the bubble is the outcome of portfolio managers' attempt to learn the model. When the bubble takes off (and when it collapses), they are learning more slowly about the model than they are changing it by revising the linear combination of chartist and fundamentalist views they incorporate in their own forecasts. But as the weight given to fundamentalists approaches zero or one, portfolio managers' estimation of the true force changing the dollar comes closer to the true one. These revisions in weights become smaller until the approximation is perfect: portfolio managers have "caught up," by changing the model more slowly than they learn. In this sense the inability of agents with prior information to bring about immediate convergence to a rational expectations equilibrium may provide a framework in which to view "bubbles" in a variety of asset markets.

Several extensions of the model in this paper would be worthwhile. First, it would be desirable to allow chartists to use a class of predictors richer than a simple random walk. They might form their forecasts of future depreciation by using ARIMA models, for example. Simple bandwagon or distributed lag expectations for chartists would be the most plausible since they capture a wide range of effects and are relatively simple analytically. Second, we might want to consider extensions which give the model local stability in the neighborhood of $= 1$. Small perturbations from equilibrium would then not instantly cause portfolio managers to begin losing faith in fundamentalist counsel. Only sufficiently large or prolonged perturbations, would upset portfolio managers' views enough to cause the exchange rate to break free of its fundamental equilibrium.

NOTES

1. Two surveys of standard asset-market models of exchange rates are Frankel (1983) and Shafer and Loopesko (1983).

2. To the extent that the monetarist model attributes the decrease in expected inflation to correct perceptions of a decreased rate of money supply growth, it could be considered as one of those mentioned above that are driven by the asset supply process. The same is true of the overshooting model. The point about asset demand versus asset supply is that rates of return are a more promising set of data with which to explain recent developments than are observed asset supplies.

3. The overshooting model, developed by Dornbusch (1976) to explain the price of foreign exchange, also has important implications for the price of agricultural commodities. Frankel (1986) presents the theoretical model in the latter context. Frankel and Hardouvelis (1985) finds empirical support for the model in the weekly Fed money announcements. Frankel (1984) offers an overview of these and other implications for commodity prices.

4. The use of the long-term real interest differential originated with Isard (1983). Other references include Shafer and Loopesko (1983) and Council of Economic Advisers (1984).

5. Sachs (1985), Hooper (1985), Hutchinson and Throop (1985) and Feldstein (1986).

6. Kling (1985) also argues that the value of the dollar rests on market expectations that do not embody a return to steady state. Ten years earlier, McKinnon (1976) attributed exchange rate volatility to a "deficiency of stabilizing speculation" that is, an unwillingness of investors to take open positions based on fundamentals equilibrium, rather than to "high capital mobility with rational expectations" as the orthodoxy has it.

7. Frankel (1985).

8. Studies regressing against the forward discount include Tryon (1979), Levich (1980), Bilson (1981), Longworth (1981), Longworth, Boothe and Clinton (1983), Fama (1984) and Huang (1984). Cumby and Obstfeld (1984) regressed against the interest differential and again found that for most exchange rates the coefficient was significantly less than 1.0 and even less than zero. These findings are also consistent with those of Meese and Rogoff (1983) that the random walk predicts not only better than other models, but better than the forward rate as well.

9. During the period June 1981 to December 1985 the 12-month forward markets were significantly biased (underpredicting the value of the dollar) even *unconditionally*. In other words, one could have made money by following the rule to be always long in dollars regardless what the forward discount was (Frankel and Froot (1986, Table 2.3)).

10. Krugman and Marris did not say that there was any reason to think that the dollar plunge would necessarily come in the next year; the focus was on the market's expected long-term rate of depreciation implicit in the long-term interest differential. We have no tests of unbiasedness going out a year or more. The problem is not the absence of a forward market going out more than a year; we can always use the long-term interest differential. The problem is rather that twelve years of floating-rate data would not offer enough independent observations.

11. The Economist survey covers 13 leading international banks and has been conducted six times a year since 1981. The American Express survey covers 250 to 300 central bankers, private bankers, corporate treasurers and economists, and has been conducted more irregularly since 1976.

12. Similarly an increase in U.S. monetary uncertainty could explain higher U.S. interest rates, but not the appreciation of the dollar. On these points, see Branson (1985) and The Council of Economic Advisers (1984, pp. 54-55).

13. One widely cited piece of evidence against the safe haven hypothesis is that the increase in U.S. real interest rates was accompanied by a lower investment rate averaged over the 1981-85 period, not a higher one. (See, for example, Friedman (1985) or Frankel (1985).) However others dispute this calculation; see Blanchard and Summers (1984). Another piece of evidence against

the safe haven hypothesis is that the correlation between U.S. stock market price changes and those abroad (Germany or Japan) has been positive; Obstfeld (1985) argues that if portfolio demands had exogenously shifted from foreign assets to U.S. assets, the U.S. stock market boom should have been accompanied by a stock market decline abroad. See also Feldstein (1986, 7-8).

14. "Small" might be defined as less than 50 basis points, to allow for differences in default risk and tax treatment attaching to the particular security, as well as inevitable minor differences in timing.

15. In 1978 the differential between the domestic U.K. and domestic U.S. interest rate is negative (columns 4 or 5 in Table 2.2). This is because of the above-mentioned U.K.-capital controls that were removed in 1979, as is evident from the differential between the Europound interest rate and domestic U.K. rates (column 2 or 3 in Table 2.2a).

16. Evans (1986) avoids this problem by employing a nonparametric sign test of the forward rate prediction errors.

17. The Money Market Services Survey has been conducted weekly or biweekly since 1983. For a more extensive analyses of this survey data set, see Dominguez (1986), Frankel and Froot (1987), and Froot and Frankel (1986).

18. In the regressions reported in Tables 2.3, 2.4 and 2.5, we use Seemingly Unrelated Squares (SUR) to exploit efficiently the contemporaneous correlation across currencies. Each currency was given its own constant term, but the constants are not reported here. See Frankel and Froot (1987) for more detail on the behavior of the survey numbers in terms of standard models of expected depreciation.

19. In Frankel and Froot (1987), we correct for the low Durbin-Watson statistics in these regressions (and those in Tables 2.4 and 2.5) using a three stages least squares estimation technique which allows for first order serial correlation in the residuals. The results are not repeated here since they are very similar to the SUR estimates already reported in Tables 2.3-2.5.

20. It sounds strange to describe 6 to 12 months as "long-term." But such descriptions are common in the foreign exchange markets.

21. Figlewski (1978, 1982) considers an economy in which private information, weighted by traders' relative wealths, is revealed in the market price.

22. See, for example, Frankel (1985).

23. Assume that prices evolve slowly according to $\pi\,(\gamma(s-p) - \sigma(i-i^*))$ (where γ and σ are the elasticities of goods demand with respect to the real exchange rate and the interest rate, respectively), that the interest rate differential is proportional to the gap between the current and long-run price levels, $\lambda\,(i-i^* + p-\bar{p}$ (where Υ is the semi-elasticity of money demand with respect to the interest rate) and that the long-run equilibrium exchange rate is given by long-run purchasing power parity, $\bar{s} = \bar{p}$. Then it can be shown that rationality implies:

$$\theta = \frac{1}{b-c} = \frac{\pi}{2\lambda}(\gamma\lambda+\sigma+(\gamma^2\lambda^2+ 2\lambda\gamma\sigma + \sigma^2+4)^{\frac{1}{2}})$$

24. In this case, however, $\omega(t)$ does not have a closed analytic form.

25. We do not consider the third case, because equations (23) and (24) are not defined at $1 + c\theta\omega(t) - \delta c = 0$.

26. The following intuition may help see why the system is stable when portfolio managers are "fast" learners and unstable when they are "slow" learners. Suppose the value of the dollar is above $\bar{s}$, so that portfolio managers are predicting depreciation at the rate $\omega\theta\,(\bar{s}-s(t))$. If the spot rate were to start depreciating at a rate slightly faster than this, portfolio managers would then shift $\omega(t)$ upwards, in favor of the fundamentalists. Under what circumstances would these hypothesized dynamics be an equilibrium? Recall from equations (21) and (22) that if δ is big, portfolio managers place substantial weight on new information. The larger is δ, the more quickly the spot rate changes. It is easy to show that if portfolio managers are fast learners (i.e., if $\delta > 1/c + \theta\omega$), they update ω so rapidly that the resulting rate of depreciation must in fact be greater than $\omega\theta\,(\bar{s}-s(t))$. Thus the system is stable. Alternatively, if portfolio managers are "slow" learners, $\delta < 1/c + \theta\omega$, they heavily discount new information and therefore change $\omega(t)$ too slowly to generate a rate of depreciation greater than $\omega\theta\,(\bar{s}-s(t))$. If we instead hypothesize an initial rate of depreciation which is less than $\omega\theta\,(\bar{s} - s(t))$, portfolio managers would tend to shift ω downwards, more towards the chartists. From equation (22), a negative $\dot{\omega}(t)$ causes the spot rate to appreciate. Thus slow learning will tend to drive the spot rate further away from the long-run equilibrium (given $0 < \omega < 1$), making the system unstable.

27. There is a second root, $\omega = -1/(\theta c)$, which we rule out since it is less than zero.

28. The assumption that ε_{t+1} exhibits such conditional heteroscedasticity results in a particularly convenient expression for δ_t (equation (A2) below). Under the assumption that ε_{t+1} is distributed normally $(0, \sigma^2)$, δ_t depends on all past values of the spot rate, $\delta_t = \tau / (\tau\theta \sum_{i=1} (\bar{s} - s_{t-i}) + T_0)$.

29. If the prior distribution is normal, the precision is equal to the reciprocal of the variance.

APPENDIX A

In this section we consider the problem which portfolio managers face: how much weight should they give to new information concerning the "true" level of $\omega(t)$. After we obtain an explicit formulation for these optimal Bayesian weights, we report their effects on the simulated path of the dollar.

Even though in the model of the spot rate given by equation (9) the value of the currency is fully deterministic, individual portfolio managers who are unable to predict accurately ex ante changes in the spot rate may view the future spot rate as random. They would then form predictions of future depreciation on the basis of observed exchange rate changes and their prior beliefs. At each point in time, portfolio managers therefore view future depreciations as the sum of their current optimal predictor and a random term:

$$\Delta s_{t+1} = \omega_t \theta(\bar{s} - s_t) + \varepsilon_{t+1} \tag{A1}$$

where ε_{t+1} is a serially uncorrelated normal random variable with mean 0 and variance $\theta(\bar{s} - s_{t-1}) / \tau$.[28] Using Bayes' rule, the coefficient ω_t may be written as a weighted average of the previous period's estimate, ω_{t-1}, and information obtained from the contemporaneous realization of the spot rate:

$$\omega_t = \frac{T_t}{T_t + \tau}\omega_{t-1} + \frac{\tau}{T_t + \tau}\left[\frac{\Delta s_t}{\theta(\bar{s} - s_{t-1})}\right] \tag{A2}$$

where $T_t = T_{t-1} + \tau$. Thus, if portfolio managers use Bayesian techniques, the weight they would give to the current period's information may be expressed as:

$$\delta_t = \tau / (\tau t + T_0) \tag{A3}$$

where T_0 is the precision of portfolio managers' prior information.[29] Equation (A3) shows that the weight which portfolio managers give to new information would fall over time as decision makers gain more confidence in their prior distribution, or as the prior distribution for the future change in the spot rate converges to the actual posterior distribution. If, however, portfolio managers suspect that the spot rate is nonstationary, past information would be discounted relative to more recent observations. Instead of combining prior information in the form of an OLS regression of actual depreciation on fundamentalist expectations (as they do above), portfolio managers might use a varying parameter technique to take into account the nonstationarity. In this case, the weight they put on new information might not decline over time to zero.

Computing δ_t using equation (A3) does not change substantially the results of the simulations presented in the text. Nevertheless the following pages contain the outcome of simulations using Bayesian δ's. Figures 5 and 6 give s(t) and ω(t) holding fundamentals constant (note that the spot rate approaches the higher equilibrium more slowly than in the comparable figures in the test, Figures 1 and 2). Figures 7 and 8 add to this changing fundamentals according to equations (26) and (27) in the text. Table 2.9 reports the simulated expectations of our three sets of agents as well as the rationally expected depreciation, comparable to Table 2.8 in the text.

FIGURE 5

SIMULATED VALUE OF THE DOLLAR ABOVE ITS LONG RUN EQUILIBRIUM

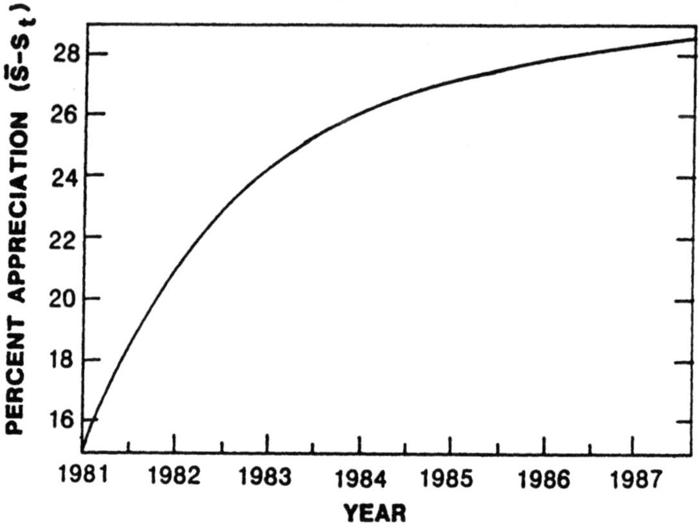

FIGURE 6

SIMULATED WEIGHT PLACED ON FUNDAMENTALIST EXPECTATIONS BY PORTFOLIO MANAGERS

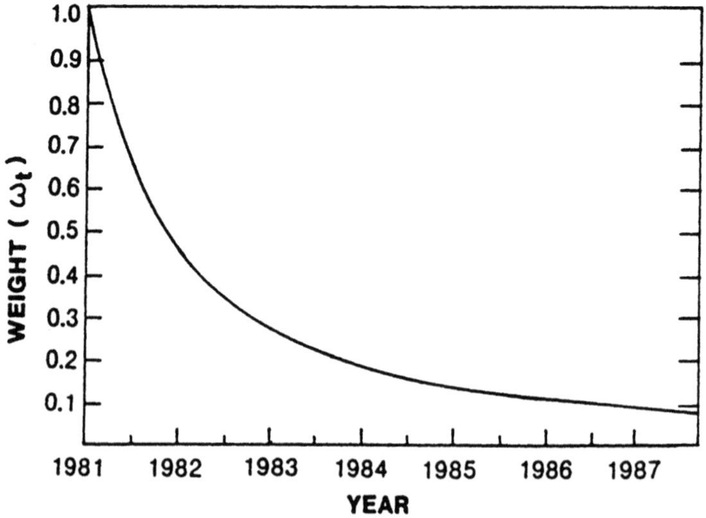

FIGURE 7

SIMULATED VALUE OF THE DOLLAR ABOVE ITS LONG RUN EQUILIBRIUM

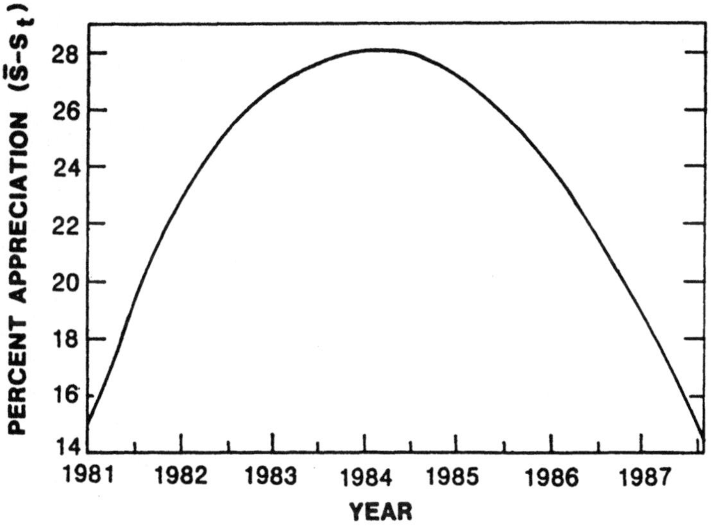

FIGURE 8

SIMULATED WEIGHT PLACED ON FUNDAMENTALIST EXPECTATIONS BY PORTFOLIO MANAGERS

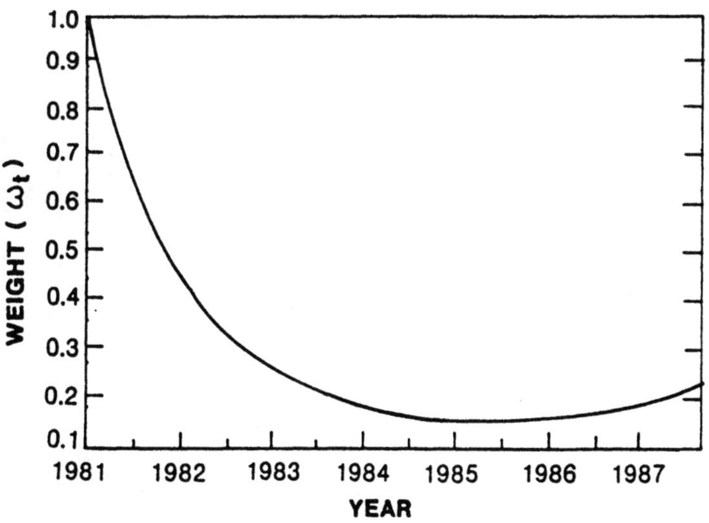

TABLE 2.9
Alternative Measures of Expected Depreciation (in percent per annum)

Expectation from:	Line	1981	1982	Year 1983	1984	1985	1986
Chartists in the simulation	(1)	0	0	0	0	0	0
Fundamentalists in the simulation	(2)	8.12	10.01	10.97	11.10	10.17	8.27
Economist 6 Month Survey	(3)	8.90	10.31	10.42	11.66	4.00	NA
Weighted Average Expected Depreciation in the Simulation	(4)	4.83	3.08	2.20	1.77	1.62	1.56
Rationally Expected Depreciation in the Simulation	(5)	-4.13	-4.45	-2.27	-0.30	2.18	4.48
Actual Forward Discount	(6)	3.74	3.01	1.10	3.07	-0.16	NA

Notes: Fundamentalists in the simulation use regressivity parameter of .045, implying that about 70% of the contemporaneous overvaluation is expected to remain after one year. The Economist 6 month survey includes data through April 1985. Weighted average expected depreciation in the simulation is a weighted average of chartists and fundamentalists, where the weights are those of portfolio managers. Rationally Expected Depreciation is the perfect foresight solution given by equations (19) and (20). The actual 6 month forward discount includes data through September 1985.

REFERENCES

Bilson, J. "The Monetary Approach to the Exchange Rate--Some Empirical Evidence," Staff Papers, International Monetary Fund. 25(March 1978).

_______. "The Speculative Efficiency Hypothesis," Journal of Business. 54(July 1981): 435-51.

_______. "Macroeconomic Stability and Flexible Exchange Rates," American Economic. Papers and Proceedings 75, 2(May 1985): 62-67.

Blanchard, O. and L. Summers. "Perspectives on High World Real Interest Rates," Brookings Papers on Economic Activity. 2(1984): 273-324.

Boothe, P., K. Clinton, A. Cote, and D. Longworth. International Asset Substitutability: Theory and Evidence for Canada. (Ottawa), Bank of Canada, February 1985.

Branson, W. H. "Causes of Appreciation and Volatility of the Dollar," in The U.S. Dollar--Recent Developments and Policy Options. A symposium sponsored by the Federal Reserve Bank of Kansas City at Jackson Hole, Wyoming, August 21-23, 1985, pp. 33-52.

Claassen, E. and C. Wyplosz. "Capital Controls: Some Principles and the French Experience," Annales de l'INSEE. 47-48 (June-December): 327-67.

Council of Economic Advisers. The Economic Report of the President. Washington, D.C., 1984.

Cumby, R. and M. Obstfeld, "International Interest Rate and Price Level Linkages under Flexible Exchange Rates: A Review of Recent Evidence," Exchange Rate Theory and Practice. J. Bilson and R. Marton (eds.), Chicago: University of Chicago Press (1984).

Dooley, M. and P. Isard. "Capital Controls, Political Risk and Deviations from Interest-Rate Parity," Journal of Political Economy. 88, 2(April 1980): 370-384.

_____. "The Appreciation of the Dollar: An Analysis of the Safe-Haven Phenomenon," International Monetary Fund. DM/85/20 (April 1985).

Dominguez, K. "Expectations Formation in the Foreign Exchange Market: New Evidence from Survey Data." Yale University, (January 1986).

Dornbusch, R. "Expectations and Exchange Rate Dynamics," Journal of Political Economy. 84(December 1976): 1161-76.

Dornbusch, R. "Equilibrium and Disequilibrium Exchange Rates," Zeitschrift fuer Wirtschafts und Sozialwissenschaften. 102, 6(1982): 573-99.

_______. Comment on Loopesko and Shafer, Brookings Papers on Economic Activity. 1(1983), pp. 78-84.

Evans, G. "Speculative Bubbles and the Sterling-Dollar Exchange Rate: A New Test," American Economic Review 76(1986).

Fama, E. "Forward and Spot Exchange Rates," Journal of Monetary Economics. 14(1984): 319-338.

Feldstein, M. "The Budget Deficit and the Dollar," NBER Working Paper, No. 1898, April 1986.

Figlewski, S. "Market 'Efficiency' in a Market with Heterogeneous Information," Journal of Political Economy. 86, 4(1978): 581-597.

________. "Information Diversity and Market Behavior," Journal of Finance. (March 1982): 87-102.

Frankel, J. "Monetary and Portfolio Balance Models of Exchange Rate Determination," Economic Interdependence and Flexible Exchange Rates. J. Bhandari, (ed.), Cambridge, M.I.T. Press, 1983.

________. The Yen/Dollar Agreement: Liberalizing Japanese Capital Markets. Policy Analyses in International Economics No. 9 (Washington, D.C.: Institute for International Economics, 1984).

________. "Commodity Prices and Money: Lessons from International Finance," American Journal of Agricultural Economics. 66(1984): 560-566.

________. "The Dazzling Dollar," Brookings Papers on Economic Activity. 1(1985): 199-217.

________. "Expectations and Commodity Price Dynamics: The Overshooting Model," American Journal of Agricultural Economics. 68(May 1986): 344-348.

Frankel, J. and K. Froot. "Using Survey Data to Test Some Standard Propositions Regarding Exchange Rate Expectations," NBER Working Paper No. 1672 (August 1985). Revised, American Economic Review 77,1(March 1987): 133-153.

Frankel, J. and G. Hardouvelis. "Commodity Prices, Money Surprises and Fed Credibility," Journal of Money, Credit and Banking. 17(1985): 425-438.

Frenkel, J. "A Monetary Approach to the Exchange Rate: Doctrinal Aspects and Empirical Evidence," Scandinavian Journal of Economics. 78(May 1976): 200-24.

Frenkel, J. and R. Levich. "Covered Interest Arbitrage: Unexploited Profits?" Journal of Political Economy. 83, 2(April 1975): 325-38.

Frenkel, J. and R. Levich. "Transaction Costs and Interest Arbitrage: Tranquil versus Turbulent Periods," Journal of Political Economy. 85, 6(December 1977): 1209-26.

Friedman, B. "Implications of the U.S. Net Capital Inflow," How Open is the U.S. Economy?, Conference at the Federal Reserve Bank of St. Louis, October 11-12, 1985.

Froot, K. and J. Frankel, "Interpreting Tests of Unbiasedness in the Forward Discount Using Surveys of Exchange Rate Expectations." University of California, Berkeley (June 1986).

Giavazzi, F. and M. Pagano. "Capital Controls and the European Monetary System," Capital Controls and Foreign Exchange Legislation. Occasional Paper, Euromobiliare, Milano (June 1985).

Hartman, D. "The International Financial Market and U.S. Interest Rates," Journal of International Money and Finance. 3(April 1984): 91-103.

Hooper, P. "International Repercussions of the U.S. Budget Deficit," Brookings Discussion Papers, No. 27, February 1985.

Hutchinson, M. and A. Throop, "U.S. Budget Deficits and the Real Value of the Dollar", Economic Review. 4, Federal Reserve Bank of San Francisco, (Fall 1985): 26-43.

Isard, P. "An Accounting Framework and Some Issues for Modeling How Exchange Rates Respond to the News," Exchange Rates and International Macroeconomics. J. Frenkel, (ed.), Chicago: University of Chicago Press (1983): 19-66.

Ito, T. "Capital Controls and Covered Interest Parity," NBER Working Paper No. 1187 (August 1983).

Kling, A. "Anticipatory Capital Flows and the Behavior of the Dollar," International Finance Discussion Paper. No. 261, Federal Reserve Board (August 1985).

Kohlhagen, S. "The Behavior of Foreign Exchange Markets--A Critical Survey of the Empirical Literature," N.Y.U. Monograph Series in Finance and Economics (1978).

Kouri, P. J. K. "The Exchange Rate and the Balance of Payments in the Short Run and the Long Run: A Monetary Approach," Scandinavian Journal of Economics. (1976): 280-304.

Kreicher, L. "Eurodollar Arbitrage," Federal Reserve Bank of New York Quarterly Review. 7, 2(Summer 1982): 10-12:

Krugman, P. "Is the Strong Dollar Sustainable?" in The U.S. Dollar: Recent Developments, Outlook, and Policy Options. A symposium sponsored by the Federal Reserve Bank of Kansas City, at Jackson Hole, Wyoming, August 21-23, 1985, pp. 103-133.

Longworth, D. "Testing the Efficiency of the Canadian-U.S. Exchange Market under the Assumption of no Risk Premium," Journal of Finance. 36(March 1981): 43-49.

Marris, S. Deficits and the Dollar: The World Economy at Risk. Policy Analyses in International Economics (Washington: Institute for International Economics, 1985).

McCormick, F. "Covered Interest Arbitrage: Unexploited Profits? Comment," Journal of Political Economy. (1979).

McKinnon, R. "Floating Exchange Rates 1973-74: The Emperor's New Clothes," Carnegie-Rochester Conference Series on Public Policy. 3(1976): 79-114.

Meese, R. and K. Rogoff, "Empirical Exchange Rate Models of the Seventies: Do They Fit Out of Sample?" Journal of International Economics. 14(February 1983): 3-24.

Mussa, M. "The Exchange Rate, the Balance of Payments and Monetary and Fiscal Policy under a Regime of Controlled Floating," Scandinavian Journal of Economics. 78(May 1976): 229-48.

Obstfeld, M. "Floating Exchange Rates: Experience and Prospects," Brookings Papers on Economics Activity. 2(1985).

Otani, I. and S. Tiwari, "Capital Controls and Interest Rate Parity: The Japanese Experience 1978-81," Staff Papers, International Monetary Fund (Washington, December 1981).

Rodriguez, C. A. "The Role of Trade Flows in Exchange Rate Determination: A Rational Expectations Approach," Journal of Political Economy. 88(1980): 1148-58.

Sachs, J. "The Dollar and the Policy Mix: 1985", Brookings Papers on Economic Activity. 1(1985): 117-185.

Shafer, J. and B. Loopesko. "Floating Exchange Rates after Ten Years." Brookings Papers on Economic Activity. 1(1983): 1-70.

Shiller, R. "Stock Prices and Social Dynamics," Brookings Papers on Economic Activity. 2(1984): 457-510.

Tryon, R. "Testing for Rational Expectations in Foreign Exchange Markets," International Finance Discussion Paper No. 139. Federal Reserve Board (1979).

3
The U.S. Price Level and Dollar Exchange Rate

Ronald McKinnon

In his 1967 presidential address to the American Economic Association, Milton Friedman usefully distinguished what monetary policy can and cannot do. The central bank cannot be expected to achieve sustained control over real variables such as output or unemployment. Nor in free financial markets can it succeed, other than temporarily, in pegging interest rates.

More positively, Friedman suggested that "of the various alternative magnitudes it (the central bank) can control, the most appealing guides for policy are exchange rates, the price level defined by some index, and the quantity of some nominal monetary total.... Of the three guides listed, the price level is clearly the most important in its own right" (page 108).

Friedman deemed exchange rates to be the least desirable guide for the United States of the 1960s. "Far better to let the market, through floating exchange rates, adjust to world conditions the 5 percent or so of our resources devoted to international trade while reserving monetary policy to promote the effective use of the other 95 percent" (page 108). This intellectual support for an inward-looking monetary policy remains dominant among American macroeconomists in the significantly more open U.S. economy of the present day.

Though differing over monetary strategy--whether exchange rates should float and whether the rate of growth in domestic money should be fixed--let us adopt Friedman's same basic

---

Ronald McKinnon, Professor of Economics, Stanford University, Stanford, CA.

objective. *The long run goal of monetary policy is to stabilize the purchasing power of the national money, while avoiding short-run cycles of inflation or deflation.*

Unfortunately, the central bank has no direct means of stabilizing broad price indices such as the GNP deflator, producer price index (PPI), or the consumer price index (CPI). Proposals have been advanced to stabilize much narrower price indices by the monetary authority directly intervening in markets for a few homogeneous primary commodities (Graham, 1942; Hart, 1976),[1] or simply promulgating such a commodity price index to be the standard of value (Hall, 1983). But these proposals turn out to be impractical or undesirable (Friedman, 1951; McKinnon, 1979).

So in practice, central banks intervene only in financial markets, in domestic bonds or foreign exchange, to determine the domestic money supply and--eventually--the prices of goods and services. And Friedman correctly identifies the fundamental problem with this indirect approach: monetary actions taken today need not affect the price level for some months or years hence. While interest rates and exchange rates in financial markets react quickly to new money issue, goods markets react sluggishly.

Consequently, information from the national income accounts on the current state of business--inflation or deflation, strength or weakness--can be a treacherous monetary indicator. Overreacting to current deflationary pressure (knowledge of which is itself only available with a considerable lag), the central bank might increase the money supply unduly and cause an excess demand for goods, price inflation and cyclical instability in the future.

The problem can be recast into one of balancing the demand for and supply of domestic money at the existing price level. How can the central bank judge when the current stock of nominal money--with ongoing and suitably controlled growth--provides just those real balances that people wish to hold into the indefinite future on the (correct) expectation that the price level won't change? Meanwhile, interest rates and possibly exchange rates change continually and quickly in response to shifting money-market conditions.

In this paper, two approaches towards signalling--and resolving this most basic problem of monetary control--are analyzed and tested empirically for the United States. The first is Professor Friedman's, what I shall call the 'domestic monetarist' position, which relies purely on domestic monetary indicators. The second takes a more open-economy approach by utilizing additional information from the dollar exchange rate and movements in foreign money supplies.

DOMESTIC MONETARISM AND THE INSULAR ECONOMY

For a financially mature economy like the United States, Friedman posits that the domestic demand for money is relatively stable and insulated from international influences. True, there may be some short-run fluctuations in liquidity preference or in interest rates which influence the demand for money. But empirically these are neither predictable nor persistent.

In the 1950s and 1960s, the American economy was large and relatively autonomous within the industrial world. Imports were not a high proportion of GNP and were confined to a fairly narrow spectrum of primary products and manufactures. The dollar was dominant as a reserve currency with a good record of price stability, and access to other country's capital markets was limited by exchange controls and by their lack of depth. Under the prevailing Bretton Woods System of fixed exchange rates, other countries generally subordinated their monetary policies to maintain roughly the same rate of price inflation in tradeable goods as that which prevailed in the United States.

Elsewhere (McKinnon, 1981), I have characterized the United States of the 1950s and 60s as an *insular economy*: one with limited financial and commodity arbitrage with the outside world, but not one fully closed to foreign trade.

In an insular economy, the central bank best confines its attention to purely domestic monetary indicators. Insofar as domestic money holders become nervous about the future course of inflation, "real" domestic assets such as goods, land, objects of art and so on are the natural inflation hedges into which they might initially shift--until nominal interest rates on domestic financial assets are bid up as a sufficient offset. Holdings of foreign exchange are not significant items on their menu of alternative assets.

Could the domestic interest rate(s) indicate when current monetary policy was too tight or too easy? Unfortunately, no, whether or not the economy is insular or open. Although immediately available, information from interest rates is ambiguous when inflationary expectations are volatile.

For example, an increase in nominal interest rates could signal an upward shift in the money-demand function--genuinely tight money, or signal that inflationary expectations have risen and bond holders are demanding higher yields to compensate for money being too plentiful. Indeed, in 1979 the U.S. Federal Reserve System stopped keying on the (Federal Funds) interest rate precisely because of this dilemma. Increasing interest rates in 1977 and 1978 induced the Fed to supply too much money, thereby contributing to the inflationary explosion of 1979-80.

To minimize having the central bank itself be a source of instability in their (implicitly) insular economy, therefore, domestic monetarists inspired by Professor Friedman (1960) would fix domestic money growth at some low level--say three to five percent per year--whatever the central bank's best guess of expected long-run growth in real output minus any projected trend in the velocity of money. The precise number chosen is less important than the forward commitment to keep money growth constant, thus providing assurance to the general public that major inflations or deflations will be avoided.

Deviations of domestic money growth from this long run target, information which may only be known some weeks later, is the "signal" domestic monetarists would have the central bank use to either tighten or loosen up. Indeed, in an insular economy with a stable demand for domestic money, changes in the domestic money supply itself should satisfactorily predict changes in domestic prices some months, or a year or two hence. And, as shown below, U.S. M1 was a sufficient statistic for predicting U.S. prices during the 1950s and 1960s--a period during which the Fed was relatively successful in stabilizing the American price level.

MONETARY CONTROL IN AN OPEN ECONOMY

But the monetary history of the United States from the early 1970s into the 1980s is quite a different story. The American economy became highly open in the following important respects:

(1) In international commodity trade, foreign price competition at the prevailing exchange rate strongly affects a very broad spectrum of American agriculture, mining, and manufacturing; and

(2) International capital flows among the industrial countries are virtually unrestricted; and

(3) While still the dominant reserve currency for denominating internationally liquid assets, the dollar now faces substantial rivalry from other hard currencies such as the yen and deutsche mark. Not only Americans, Japanese, and Europeans, but portfolio managers

in other countries (LDCs) continually shift their asset preferences--mainly for interest-bearing bonds--among dollars, yen, and various European currencies.

But why might this increased openness make any difference to the problem of monetary stabilization? After all, the supply of U.S. money remains, as in an insular economy, the dominant control variable for influencing the American price level--with uncertain lags and credibility problems in linking future expectations to present policy.

However, if expectations of future American price inflation should change, liquid foreign exchange assets--and not domestic inventories of goods or other physical assets--are now the preferred portfolio alternative to holding dollar claims. At the margin, switching to foreign bonds or bank accounts--denominated in freely convertible hard currencies such as marks or yen--is now much more convenient than acquiring relatively illiquid physical assets.

Under floating exchange rates in the 1970s and 80s, foreign central banks are no longer officially obligated to subordinate their monetary policies to that being followed by the United States. Thus foreign hard currencies are more differentiated from dollars with respect to potential inflation or deflation in the future. Of course, most central banks will claim that they intend to stabilize their domestic price level. But gimlet-eyed international investors will inevitably suspect that some are being more successful than others--given the very difficult problem of intertemporal monetary control which they all face.

This relationship is symmetrical: investors holding yen or mark assets can easily switch into dollars should they change their assessments of Japanese or German monetary policies, wealth taxes, or other sources of future risk. In an open economy without exchange controls, purely domestic inflation hedges become less attractive than they would be in an insular economy.

Increased openness is, therefore, at once a disadvantage and an advantage for resolving our intertemporal problem of monetary control.

On the one hand, the option to acquire liquid foreign exchange makes holders of dollar assets much more sensitive to changing assessments of American monetary policy--as well as future taxation and political risk--relative to similar policies in other countries. The effective demand for both money and bonds denominated in dollars has become more volatile--potentially complicating the Federal Reserve's intertemporal problem of monetary control. The unexpectedly sharp two-to-four-year cycles of inflation in the 1970s, and deflation in the early 1980s, are the unfortunate consequences.

On the other hand, the foreign exchanges provide information: they immediately register pressure for or against dollar denominated assets. In particular, when exchange rates are not fixed, the floating dollar could signal the Fed when expectations of future American price inflation, and the effective demand for U.S. money, were changing. But this remains to be demonstrated.

THE ASSET APPROACH TO EXCHANGE RATES AND THE DOMESTIC PRICE LEVEL

In empirical work beginning in the mid 1970s, Jacob Frenkel, Michael Mussa and Richard Levich have established that a freely floating exchange rate behaves like an asset price. (For a recent summary, see Frenkel and Mussa (1985), and Levich (1985)). Rather than being dominated by current or past flows of imports and exports, the exchange rate varies continually to maintain day-to-day balance across international asset portfolios.

Like the prices of common stocks, the exchange rate seems to be a forward-looking variable that responds only to new information, "news", as if international investors were continually trying to anticipate what monetary and other financial policies each country might follow--or how its terms of trade might change. If some political economic event, say an election, causes people to believe that a country's relative inflation rate or other taxation of wealth will be higher in the future, its exchange rate will depreciate immediately and thus add to the inflationary pressure.

Hence, this now commonly accepted asset approach to exchange rate determination is consistent with our open-economy theory of inflation hedges sketched above. Should inflationary expectations change, foreign exchange rather than "real" assets is the most convenient alternative to holding financial assets--including money--denominated in domestic currency.

A corollary to this asset approach is that current exchange rate movements do not reflect past changes in the income, prices, or other trade and financial variables. Indeed, attempts to predict exchange rates econometrically, on the basis of generally available information from the past, have all failed out of sample (Meese and Rogoff, 1983). Given our particular concern with intertemporal monetary control, one should note that fluctuations in current exchange rates have not been explainable by (past) growth in domestic money and do indeed reflect new information beyond what past or current growth in M1, M2, the monetary base, and so on might suggest.

A further corollary is that exchange rates move further, and much more rapidly, than (equilibrium) movements in relative national price levels, trade balances, output, and so on--the "overshooting" phenomenon (Dornbusch, 1976). Even though economies are now very open, domestic price levels remain sticky when measured in the national currency. Over months and up to a year or two, fluctuations in nominal exchange rates are equivalent to changes in real rates.

On a purely statistical basis, therefore, floating exchange rates seem to lead prices rather than the other way around (Frenkel, 1978). Given the continual changes in people's perceptions of the future as revealed in their shifting international asset preferences, a floating exchange rate is nearly always "out of equilibrium" from the point of commodity markets (Ohno, 1985)--thus imposing either inflationary or deflationary pressure on the domestic price level.

THE ASYMMETRICAL POSITION OF THE UNITED STATES

While accurate as far as it goes, the asset approach treats all countries symmetrically by not differentiating among them. And in a symmetrical float, any one exchange rate might be an ambiguous monetary indicator of inflation or deflation to come within the domestic economy.[2]

First is the question of choosing that exchange rate--yen, marks, francs, guilders, lire or some combination--to which the Fed should respond. Which "hard" foreign currency should be the standard of reference?

Secondly, if just one exchange rate was considered, say the mark/dollar rate, wouldn't this reflect disturbances in the German money market as much as the American? Couldn't a rise in the mark/dollar rate simply reflect actual or expected excess money issue in Germany, and thus throw out a confusing signal for what the Fed should be doing?[3]

Fortunately, both of these potential ambiguities can be resolved by appealing to our historical experience with floating, and by noting the asymmetrical position of the United States under the continuing world dollar standard (Kenen, 1983 and McKinnon, 1979). Since the early 1970s, the dollar shows very high variance against all other currencies viewed collectively--see Figure 1 for the IMF's "merm" weighted dollar exchange rate against 17 other industrial countries.

In the two-to four year swings with which we are concerned, the dollar exchange rates of countries outside of North America are highly correlated with one another: rising together in 1971-73 and in 1977-79 and then falling sharply from 1980 through 1984--see Figures 3 and 4 for Germany, the Netherlands, the U.K. and Japan. To be sure, there are differing long-term trends in exchange rates over the past 15 years: with the yen and mark tending to appreciate against the dollar, and sterling (as well as French francs and lire) tending to depreciate. Nevertheless, the European and Japanese dollar exchange rates have tended to move similarly on a quarterly or annual basis.[4]

The upshot is that shifts in portfolio preferences for or against U.S. dollar assets seem to be dominated by changing expectations of what American monetary and financial policies--or commercial prospects--will be in the future. Or, putting this proposition the other way around, there is no other sufficiently large country in the system whose domestic financial disturbances--either actual or anticipated--significantly impinge on the average dollar exchange rate of the United States.

This fundamental asymmetry in the world's exchange rate system goes beyond the disproportionate economic size of the American economy. The dollar remains the vehicle currency for international capital flows and for denominating most official exchange reserves, as well as being the invoice currency for most international trade in primary commodities such as oil. Thus other countries' governments "have a view" of what their dollar exchange rates should be, and react to smooth (not very successfully) major fluctuations. When the dollar is weak they tend to expand, and when the dollar is strong they tend to contract.

Based on smoothed (5-quarter moving averages) and unsmoothed quarterly data, Figure 2 shows the strong inverse correlation between changes in the strength of the dollar in the foreign exchanges and money growth in the "rest of the world" (ROW): the 10 major industrial countries outside of the United States. Table 3.1 (based on annual data) shows how the series on ROW money growth was constructed using fixed GNP weights for the mid year (1977) in the series. Figure 2 shows particularly high foreign money growth in 1971-73 and 1977-79 when the dollar was falling, and the fall off in ROW money growth over 1980-82 when the dollar exchange rate recovered. Most recently, the rise of the dollar in 1984 forced a

FIGURE 1
U.S. EFFECTIVE EXCHANGE RATE AND
THE REST OF THE WORLD MONEY
(quarterly observations)

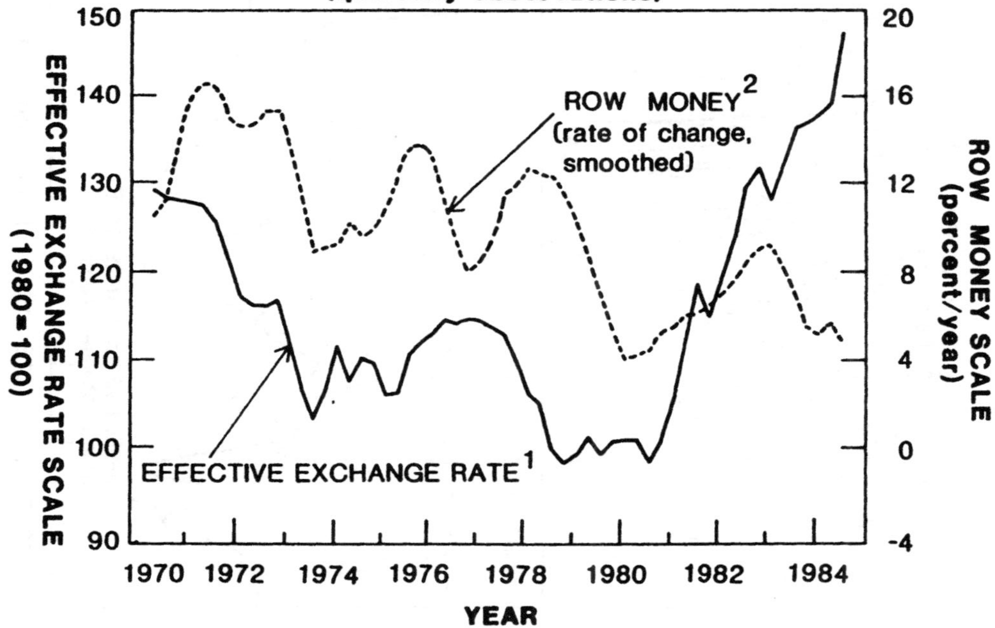

[1] IMF defination: MERM (trade) weighted nominal rate against 17 countries.
[2] Percent growth in nominal money in 10 industrial countries other than U.S.: see Table 1.

FIGURE 2
U.S. EFFECTIVE EXCHANGE RATE AND THE REST OF THE WORLD MONEY

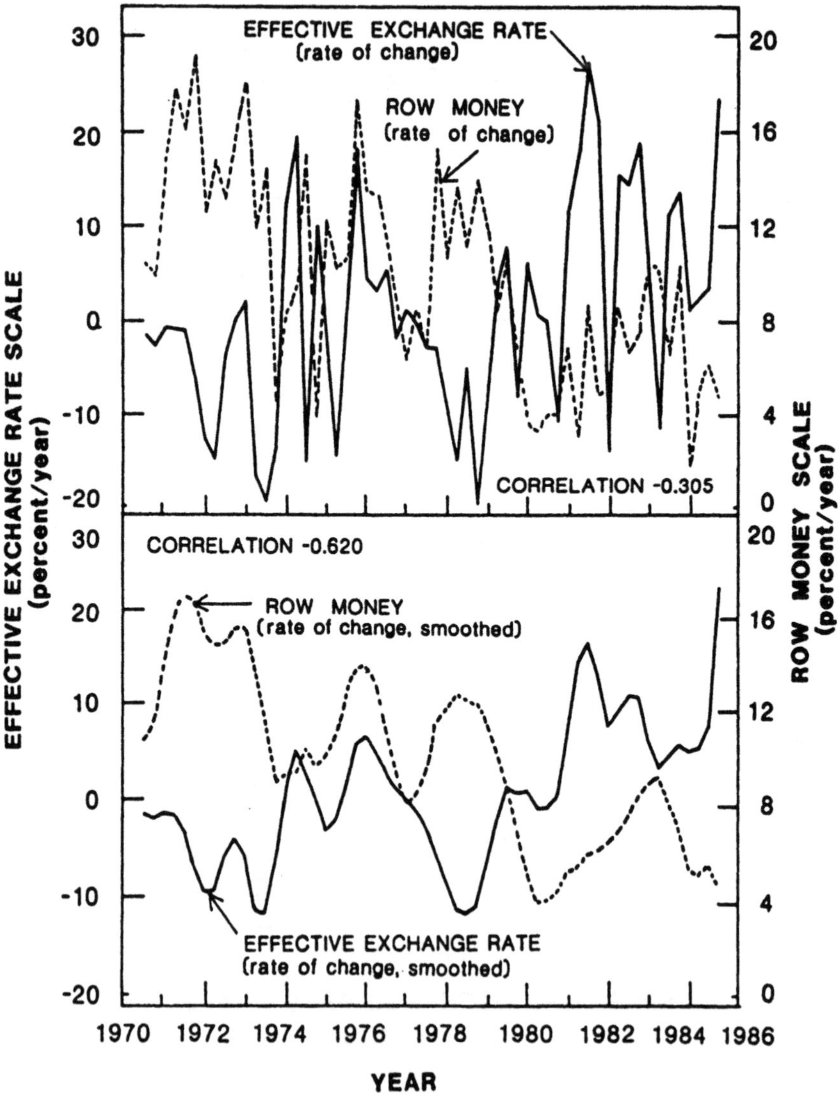

FIGURE 3
RELATIONSHIPS OF DOLLAR EXCHANGE RATES

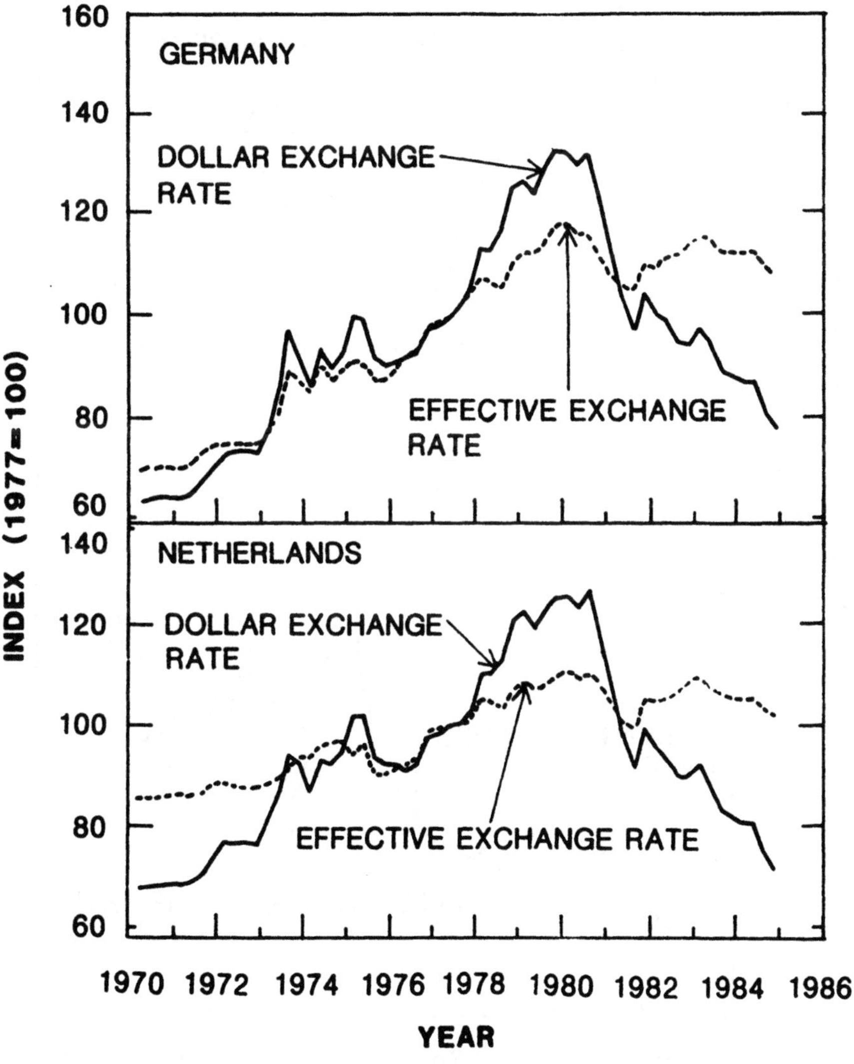

FIGURE 4
RELATIONSHIPS OF DOLLAR EXCHANGE RATES

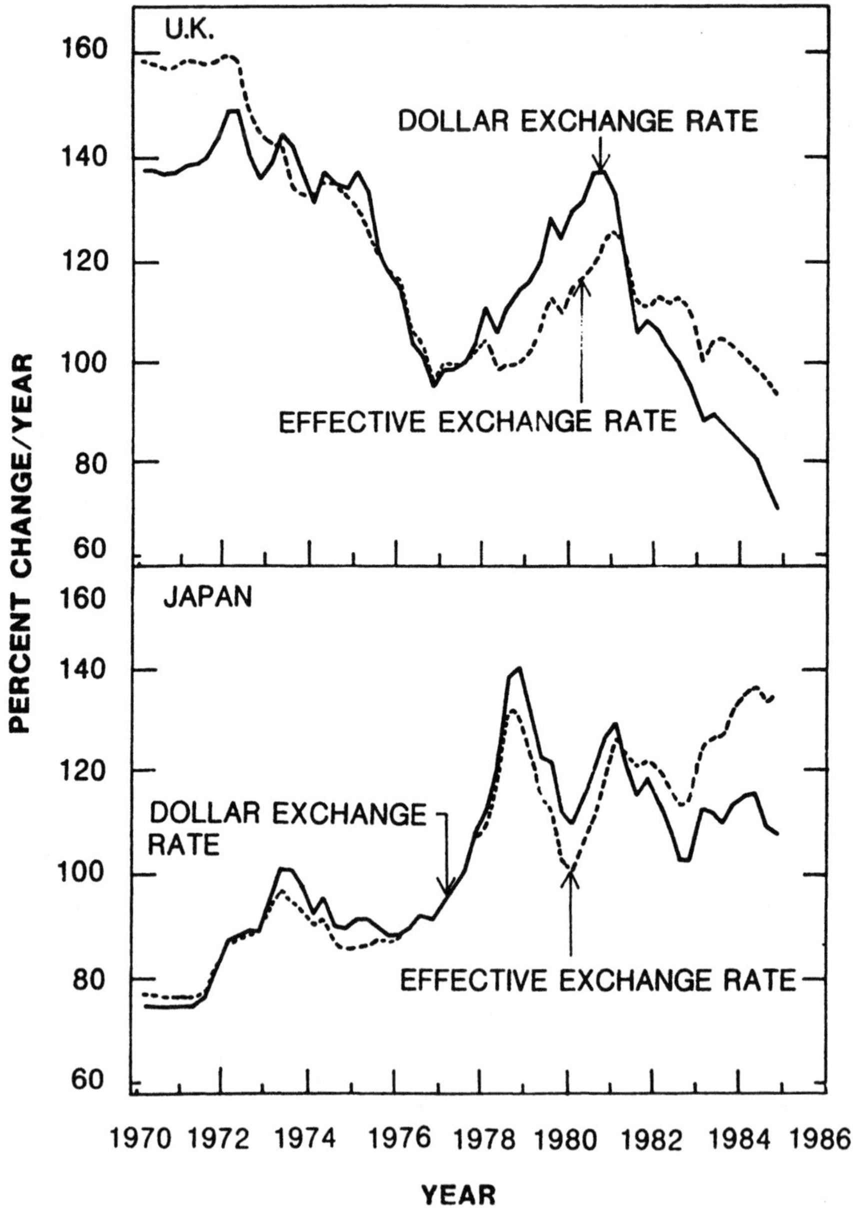

TABLE 3.1
Money Growth in Domestic Currencies, 11 Industrial Countries
(percentage change in annual averages of M1)

	Belgium	Canada	France	Germany	Italy	Japan
(Weights: GNP 1964)	(.0132)	(.0394)	(.0778)	(.0892)	(.0494)	(.0681)
1956	2.9	-1.2	10.3	7.2	8.5	16.4
1957	-0.1	4.0	8.6	12.1	6.3	4.1
1958	5.8	12.8	6.4	13.1	9.9	12.8
1959	3.2	-3.2	11.4	11.8	14.0	16.5
1960	1.9	5.1	13.0	6.8	13.5	19.1
1961	7.7	12.4	15.5	14.8	15.7	19.0
1962	7.2	3.3	18.1	6.6	18.6	17.1
1963	9.8	5.9	16.7	7.4	16.9	26.3
1964	5.6	5.1	10.3	8.3	6.7	16.8
1965	7.4	6.3	9.0	8.9	13.4	16.8
1966	6.7	7.0	8.9	4.5	15.1	16.3
1967	4.7	9.5	6.2	3.3	13.6	13.4
1968	6.8	4.4	5.5	7.6	13.4	14.6
1969	2.3	6.9	6.1	8.2	15.0	18.4
1970	-2.5	2.4	-1.3	6.4	21.7	18.3
(Weights: GNP 1977)	(.0172)	(.0487)	(.1122)	(.1122)	(.0471)	(.1404)
1971	10.3	12.7	13.7	12.0	22.9	25.5
1972	15.0	14.3	13.0	13.6	18.0	22.0
1973	9.8	14.5	9.9	5.8	21.1	26.2
1974	6.8	9.3	12.6	6.0	16.6	13.1
1975	12.4	13.8	9.9	13.8	8.3	10.3
1976	9.6	8.0	15.0	10.4	20.5	14.2
1977	8.0	8.4	7.5	8.3	19.8	7.0
1978	6.7	10.0	11.2	13.4	23.7	10.8
1979	3.5	6.9	12.2	7.4	23.9	9.9
1980	-0.2	6.3	8.0	2.4	15.9	0.8
1981	3.6	4.3	12.3	1.2	11.1	3.7
1982	3.4	2.0	14.9	3.5	9.9	7.1
1983	5.0	10.2	12.1	10.3	17.3	0
1984	3.3	2.3	8.2[b]	3.3	8.4[b]	2.3

(continued)

TABLE 3.1 (cont.)

	Nether-lands	Sweden	Switzer-land	United Kingdom	United States	World Average	Rest of World[a]
(Weights: GNP 1964)	(.0144)	(0.167)	(.0113)	(.0796)	(.5408)		
1956	-3.7	7.4	6.0	1.0	1.1	3.78	6.94
1957	-2.0	3.4	1.8	2.7	-0.6	2.43	6.01
1958	11.9	1.6	9.2	3.0	4.3	6.47	9.04
1959	4.5	18.0	6.1	4.6	0.1	4.53	9.74
1960	6.7	-1.2	10.2	-0.8	-0.4	3.72	8.58
1961	7.7	10.7	8.1	3.2	2.9	7.39	12.68
1962	7.5	5.6	16.6	4.4	2.1	6.18	10.99
1963	9.8	8.1	8.9	0.3	2.8	6.86	11.65
1964	8.5	7.7	0.2	5.0	4.1	6.16	8.59
1965	10.9	6.4	12.8	2.7	4.3	6.59	9.30
1966	7.2	9.9	3.1	2.6	4.6	6.31	8.33
1967	7.0	9.8	6.0	3.2	3.9	5.49	7.37
1968	8.8	-1.8	11.5	6.0	7.0	7.51	8.12
1969	9.4	2.0	9.5	0.4	5.9	7.00	8.30
1970	10.6	7.3	9.8	6.4	3.8	5.80	8.15
(Weights: GNP 1977)	(.0228)	(0.195)	(.0148)	(.0572)	(.4316)		
1971	16.7	9.0	18.2	11.8	6.8	12.45	16.71
1972	17.7	11.8	13.4	13.1	7.1	12.21	16.10
1973	7.4	9.6	-1.0	8.6	7.3	11.06	13.91
1974	3.1	16.3	-1.7	4.8	5.0	7.78	9.88
1975	18.7	15.2	2.4	15.6	4.7	8.83	11.96
1976	11.8	14.0	7.3	13.8	5.7	9.91	13.10
1977	14.3	8.3	4.7	14.4	7.6	8.72	9.57
1978	5.3	13.6	12.7	20.1	8.2	10.99	13.11
1979	2.7	12.7	7.8	11.5	7.7	9.23	10.39
1980	4.2	21.1	-5.4	4.9	6.2	5.53	5.01
1981	2.6	12.0	-0.9	10.0	7.2	6.50	5.96
1982	4.9	9.8	3.1	8.3	6.5	6.96	7.31
1983	10.6	11.4	7.6	13.4	11.1	10.1	9.48
1984	4.1	2.4[b]	2.5[b]	14.9[b]	6.9	6.08	5.45

Source: Federal Reserve Bank of St. Louis, "International Economic conditions," June and August 1985.

[a]United States excluded

[b]Preliminary.

reduction in ROW money growth sharply below its long-run norm (Table 3.1).

One important implication of this asymmetry is that the United States, as the center country, has had more complete independence in choosing its own monetary policies than other industrial countries--and its cycles of inflation or deflation tend to spread out into the rest of the industrial world (McKinnon, 1982). Because other countries' monetary policies are somewhat more (although by no means completely) endogenized, fluctuations in the dollar exchange rate are more likely to reflect changing money-market conditions in the United States leading to eventual world-wide inflation or deflation.

For example, if the dollar suddenly appreciates, this indicates that U.S. monetary policy has become tighter--because of an unexpected supply constraint (such as a fall in the American money multiplier) or because the effective national and international demand for U.S. money has increased. (The simple statistical regression model presented below attempts to distinguish between these two cases.) In either event, the consequential deflationary pressure on the American economy is reinforced by monetary contraction abroad and further deflation in the prices of internationally tradeable goods. These international repercussions strengthen the effect of the dollar exchange rate in predicting future American price inflation.

AMERICAN PRICES AND U.S. MONEY GROWTH: STATISTICAL EVIDENCE FROM THE 1950S AND 60S

How well does the principle of domestic monetarism, which treats the United States as if it was an insular economy, fare in the fixed exchange rate period of the 1950s and 60s in comparison to floating rates in the 1970s and 80s?

Constructing a complete structural model of the American macro economy--in which output, prices, interest rates and so on are jointly determined--is beyond the scope of this paper. Instead consider a single reduced-form regression of current U.S. price inflation on (current and) past percentage changes in U.S. narrow money--M1 as presently defined by the Federal Reserve's Board of Governors.

$$\dot{P} = C + a\dot{M} + a_{-1}\dot{M}^{US}_{-1} + \dots a_{-n}\dot{M}^{US}_{-n} + u \qquad (1)$$

The dot over each variable indicates percentage rates of change in either annual or quarterly data. Each regression is based on first differences of the logarithms of some general price index and of the money supply lagged n periods. Although perhaps losing some information contained in levels of the variables, the first difference approach has the advantage of suppressing (spurious) correlation associated with trends in which both the price level and money supply increase through time. Because n extends up to three years (12 quarters), then the 'a' coefficients pick up the impact of variance in money growth on (cyclical) fluctuations in prices.

The other major statistical problem is to choose an appropriate time period--weeks, months, quarters, years--over which to average each observation on P and M. Even if price data were available on a weekly basis, it would not be usable given the sluggishness with which the price level adjusts to any (unexpected) changes in the money supply: there would be too much serial correlation in the 'u' disturbances--as well as errors in measurement of both P and M. On the other hand, annual observations would seem to smooth too much. Intra-year cyclical fluctuations in M and P would be averaged out, leaving too few observations.

Consequently, I have chosen to run the ordinary least squares regression explaining movements in the U.S. wholesale price index (WPI) and in the GNP deflator (DEF) on an annual basis--Tables 3.3 and 3.4, and on a quarterly basis--Tables 3.5, 3.6, and 3.7. Fortunately, they tell the same interesting and sharply-defined story.

The U.S. WPI, as calculated by the International Monetary Fund, is a rather broad price index for tradeable goods including both finished manufactures and crude materials--whereas the closely related U.S. producer price index includes only finished goods. On the other hand, the GNP deflator is yet more general: including a high volume of nontradeable services whose prices move more sluggishly. Thus the WPI shows much more variance (Table 3.2) than the GNP deflator.

Nevertheless, the 1950s and 1960s, U.S. M1 explains movements in both American price indices rather well--despite all the limitations of our single-equation regression approach. The annual regressions, based on 12 observations for 1958-69, show the best $\bar{R}^2$ to be 0.47 for the WPI (Table 3.3), and to be 0.70 for the GNP deflator (Table 3.4). The signs of the 'a' coefficients are correct (positive) and add up to about .65, although the number of observations is too few to say much about levels of significance for individual coefficients.

TABLE 3.2
Price Inflation in Tradeable Goods, 11 Industrial Countries
(percentage change in annual averages of WPIs)

	Belgium	Canada	France	Germany	Italy	Japan
(Weights: GNP 1964)	(.0132)	(.0394)	(.0778)	(.0892)	(.0494)	(.0681)
1958	-4.4	0.4	5.1	-0.5	-1.7	-6.5
1959	-0.3	0.8	7.2	-0.8	-2.9	0.9
1960	1.2	0.2	3.5	1.3	0.8	1.1
1961	-0.2	0.2	3.0	1.5	0.0	1.1
1962	0.8	1.1	0.6	0.9	3.2	-1.6
1963	2.5	1.3	2.9	0.5	5.3	1.6
1964	4.7	0.9	3.5	1.0	3.0	0.4
1965	1.1	1.3	0.7	2.5	1.8	0.7
1966	2.1	2.9	2.8	1.7	1.5	2.4
1967	-0.9	1.9	-0.9	-1.0	-0.2	1.7
1968	0.2	2.2	-1.7	-0.7	0.6	1.0
1969	5.0	3.7	10.7	1.9	3.6	2.0
1970	4.7	2.4	7.5	5.0	7.4	3.7
(Weights: GNP 1977)	(.0172)	(.0487)	(.0885)	(.1122)	(.0471)	(.1404)
1971	-0.5	2.0	2.1	4.3	3.3	-0.
1972	4.0	4.3	4.7	2.5	4.1	0.8
1973	12.4	11.2	14.7	6.6	17.2	15.8
1974	16.8	19.1	29.1	13.5	40.8	31.4
1975	1.2	11.2	-5.7	4.6	8.5	3.0
1976	7.1	5.1	7.4	3.7	23.8	5.0
1977	2.4	7.9	5.6	2.7	16.6	1.9
1978	-1.9	9.3	4.3	1.2	8.4	-2.5
1979	6.3	14.4	13.3	4.8	15.5	7.3
1980	5.8	13.5	8.8	7.5	20.1	17.8
1981	8.2	10.1	11.0	7.7	16.6	1.7
1982	7.7	6.0	11.1	5.8	13.9	1.8
1983	5.2	3.5	11.0	1.5	10.5	-2.2
1984	7.4	4.1	13.3	2.9	10.4	-0.2

(continued)

TABLE 3.2 (cont.)

	Netherlands	Sweden	Switzerland	United Kingdom	United States	World Average	Rest of World[a]
(Weights: GNP 1964)	(.0144)	(.0167)	(0.113)	(.0796)	(.5408)		
1958	-1.3	4.3	-3.2	0.8	1.5	0.68	-0.30
1959	0.2	0.9	-1.6	0.3	0.2	0.57	1.00
1960	0.0	4.1	0.6	1.3	0.2	0.81	1.54
1961	-0.2	2.2	0.2	2.6	-0.4	0.47	1.50
1962	0.3	4.7	3.3	2.3	0.2	0.64	1.16
1963	2.4	2.9	3.9	1.0	-0.4	0.72	2.03
1964	6.1	3.4	1.3	3.1	0.2	1.15	2.27
1965	3.0	5.2	0.6	3.5	2.0	1.98	1.95
1966	5.0	6.4	1.9	2.9	3.4	3.02	2.57
1967	1.0	4.3	0.3	3.1	0.2	0.45	0.75
1968	1.9	2.0	0.1	4.1	2.4	1.68	0.83
1969	-2.5	3.5	2.8	3.7	3.9	3.99	4.09
1970	4.6	6.8	4.2	7.1	3.6	4.54	5.65
(Weights: GNP 1977)	(.0228)	(.0195)	(.0148)	(.0572)	(.4316)		
1971	4.5	3.2	2.1	9.1	3.3	2.94	2.67
1972	5.1	4.6	3.6	5.3	4.4	3.74	3.24
1973	6.9	10.3	10.7	7.4	13.1	12.42	11.91
1974	9.6	25.3	16.2	22.6	18.8	22.00	24.43
1975	6.7	6.4	-2.3	22.2	9.3	6.93	5.12
1976	7.8	9.0	-0.7	17.3	4.6	6.58	8.09
1977	5.8	9.2	0.3	19.8	6.1	6.35	6.55
1978	1.3	7.6	-3.4	9.1	7.8	4.99	2.86
1979	2.7	12.5	3.8	12.2	12.5	10.73	9.39
1980	8.2	13.9	5.1	16.3	14.0	13.33	12.82
1981	9.2	11.6	5.8	10.6	9.0	8.50	8.13
1982	6.6	12.6	2.6	8.6	2.1	4.80	6.85
1983	1.8	11.2	0.5	5.5	1.3[b]	2.73	3.82
1984	4.2	7.9	3.3	6.2	2.4	3.98	5.18

Source: IMF, International Financial Statistics, 1984 Yearbook and July 1985, line 63, wholesale price indices including finished goods and primary products.

[a]United States excluded.

[b]Preliminary.

TABLE 3.3
American Tradeable Goods Prices (WPI), the Dollar Exchange Rate, and Growth in U.S. and World Money (annual data, yearly averages; t-statistics in parentheses)

Fixed Exchange Rates: 1958-69

$\dot{WPI}^{US}$	C	$\dot{M}^{US}$	$\dot{M}^{US}_{-1}$	$\dot{M}^{US}_{-2}$	$\bar{R}^2$	SER	DW	Regression Method
	-0.94 (-1.35)	0.32 (1.73)	0.21 (1.29)	0.13 (0.64)	0.43	1.12	1.87	OLS
	0.83 (-1.27)	0.37 (2.18)	0.23 (1.49)		0.47	1.09	1.99	OLS
$\dot{WPI}^{US}$	C	$\dot{M}^{W}$	$\dot{M}^{W}_{-1}$	$\dot{M}^{W}_{-2}$				
	-3.14 (-0.92)	0.45 (1.04)	0.21 (0.66)	0.05 (0.15)	0.00	1.59	1.20	OLS
	-2.99 (-0.97)	0.46 (1.14)	0.22 (0.74)		0.00	1.50	1.21	OLS
	-4.10 (-1.02)	0.49 (1.22)	0.36 (1.04)		-	1.46	1.79	AR(1) (ρ=.33)

(continued)

TABLE 3.3 (cont.)

Floating Exchange Rates: 1973-84

$\dot{DEF}^{US}$	C	$\dot{M}^{US}$	$\dot{M}^{US}_{-1}$	$\dot{M}^{US}_{-2}$	$\dot{E}_{-1}$	$\dot{E}_{-2}$	R^2	SER	DW	Regression Method
	5.56 (1.52)	-0.56 (-2.00)	-0.39 (-1.35)	1.20 (2.69)			0.47	1.56	1.02	OLS
	5.75 (1.34)	-0.55 (-2.01)	-0.34 (-1.10)	1.10 (2.26)			-	1.53	1.65	AR(1) (ρ=0.44)
	1.94 (0.53)		-0.47 (-1.44)	1.25 (2.43)			0.30	1.80	1.40	OLS
	3.41 (0.99)		-0.46 (-1.53)	1.03 (2.12)	-0.12 (-1.73)		0.43	1.63	1.84	OLS
	3.35 (1.19)		-0.09 (-0.30)	0.64 (1.49)	-0.04 (-0.61)	-0.19 (-2.22)	0.61	1.33	1.52	OLS
	3.08 (1.23)			0.59 (1.59)	-0.04 (-0.58)	-0.20 (-3.06)	0.66	1.25	1.56	OLS
	2.88 (1.27)			0.61 (1.74)		-0.22 (-3.95)	0.68	1.21	1.70	OLS

-12.72 (-3.01)	0.30 (0.62)	2.00 (4.58)			0.74	2.79	2.02	OLS
-0.85 (-0.13)	-0.21 (-0.46)	1.26 (2.56)	-0.40 (-2.27)		0.82	2.30	2.33	OLS
1.63 (0.28)	0.08 (0.17)	0.70 (1.32)	-0.36 (-2.26)	-0.24 (-1.78)	0.86	2.04	1.92	OLS
2.15 (0.47)		0.72 (1.48)	-0.37 (-2.92)	-0.23 (-1.97)	0.88	1.91	1.96	OLS

Notes: $\dot{WPI}^{US}$ is percentage inflation in US Wholesale Prices, including finished goods and raw materials, as tabulated on line 63 of the IMF International Financial Statistics (IFS). $\dot{M}^{US}$ is the percentage increase in USMI (narrow money). $\dot{M}^{W}$ is the percentage increase in M1 of 11 major industrial countries using fixed GNP weights and compiled from International Economic Conditions, Federal Reserve Bank of St. Louis--see Table 1. $\dot{E}$ is percentage increase in the trade (merm) weighted value of the dollar against currencies of major US industrial trading partners; see line amx of the IFS. $\bar{R}^2$: percentage of variance explained adjusted for degrees of freedom. SER: standard error of the regression. DW: Durbin-Watson Statistic. OLS: ordinary least squares. AR(1): OLS corrected for serial correlation. The regression period reflects the span of the dependent variable. Hence 1973-84 and 1958-69 each consist of 12 annual observations.

TABLE 3.4
American GNP Deflator, the Dollar Exchange Rate, and Growth in U.S. and World Money (annual data, yearly average; t-statistics in parentheses)

Fixed Exchange Rates: 1958-69

$\dot{DEF}^{US}$	C	$\dot{M}^{US}$	$\dot{M}^{US}_{-1}$	$\dot{M}^{US}_{-2}$	$\bar{R}^2$	SER	DW	Regression Method
	2.91 (1.55)	0.16 (1.42)	0.33 (3.27)	0.15 (1.22)	0.70	0.69	0.63	OLS
	0.65 (1.53)	0.21 (1.95)	0.36 (3.47)		0.69	0.71	0.95	OLS
	0.59 (1.02)	0.19 (1.84)	0.38 (3.80)		-	0.67	1.64	AR(1) (ρ=0.31)
$\dot{DEF}^{US}$	C	$\dot{M}^{W}$	$\dot{M}^{W}_{-1}$	$\dot{M}^{W}_{-2}$				
	-1.48 (-0.54)	0.28 (0.79)	0.36 (1.37)	0.03 (0.10)	0.00	1.28	0.36	OLS
	-1.41 (-0.56)	0.29 (0.87)	0.36 (1.50)		0.08	1.22	0.38	OLS
	3.09 (0.41)	-0.09 (-0.50)	0.21 (1.18)		-	0.87	1.42	AR(1) (ρ=.93)

Floating Exchange Rates: 1973-84

$\dot{DEF}^{US}$	C	$\dot{M}^{US}$	$\dot{M}^{US}_{-1}$	$\dot{M}^{US}_{-2}$	$\dot{E}_{-1}$	$\dot{E}_{-2}$	R^2	SER	DW	Regression Method
	5.56 (1.52)	-0.56 (-2.00)	-0.39 (-1.35)	1.20 (2.69)			0.47	1.56	1.02	OLS
	5.75 (1.34)	-0.55 (-2.01)	-0.34 (-1.10)	1.10 (2.26)			-	1.53	1.65	AR(1) (ρ=0.44)
	1.94 (0.53)		-0.47 (-1.44)	1.25 (2.43)			0.30	1.80	1.40	OLS
	3.41 (0.99)		-0.46 (-1.53)	1.03 (2.12)	-0.12 (-1.73)		0.43	1.63	1.84	OLS
	3.35 (1.19)		-0.09 (-0.30)	0.64 (1.49)	-0.04 (-0.61)	-0.19 (-2.22)	0.61	1.33	1.52	OLS
	3.08 (1.23)			0.59 (1.59)	-0.04 (-0.58)	-0.20 (-3.06)	0.66	1.25	1.56	OLS
	2.88 (1.27)			0.61 (1.74)		-0.22 (-3.95)	0.68	1.21	1.70	OLS

(continued)

TABLE 3.4 (cont.)

$\dot{DEF}^{US}$	C	$\dot{M}^W$	$\dot{M}^W_{-1}$	$\dot{M}^W_{-2}$	$\dot{M}^W_{-3}$	$\dot{E}_{-1}$	$\dot{E}_{-2}$	$\bar{R}^2$	SER	DW	Regression Method
	4.38 (1.85)		-0.57 (-2.10)	0.84 (3.41)				0.47	1.56	1.60	OLS
	0.45 (0.17)		-0.33 (-1.37)	0.59 (2.66)	0.44 (2.41)			0.66	1.26	1.79	OLS
	2.99 (0.61)			0.09 (0.25)	0.34 (1.42)	-0.27 (-0.29)	-0.14 (-1.30)	0.61	1.34	1.20	OLS
	4.01 (2.14)				0.33 (1.62)		-0.18 (-2.64)	0.67	1.23	1.44	OLS

Notes: $\dot{DEF}^{US}$ is annual percentage change in the United States GNP deflator: the most general American measure of price inflation, including both goods and services.

For other definitions, see notes to Table 3. 1958-69 and 1973-74 each consist of 12 annual observations.

TABLE 3.5
American Prices, the Dollar Exchange Rate and U.S. Money Growth: Historical Comparisons (quarterly data, t-statistics in parentheses)

Dependent Variable	$\dot{M}^{US}$	$\dot{E}^{US}$	$\bar{R}^2$	SER (Percentage points)	DW	Time Period
$\dot{DEF}^{US}$	0.98 (8.24)		0.61	0.26	2.03	62.2-73.1
$\dot{WPI}^{US}$	1.62 (5.58)		0.47	0.64	2.07	62.2-73.1
$\dot{DEF}^{US}$	0.44 (1.12)		0.11	0.58	0.78	73.2-84.4
$\dot{WPI}^{US}$	0.81 (0.70)		-0.04	1.73	0.98	73.2-84.4
$\dot{DEF}^{US}$	0.57 (1.91)	-0.34 (-4.87)	0.55	0.41	1.33	73.2-84.4
$\dot{WPI}^{US}$	1.20 (1.35)	-1.07 (-5.17)	0.49	1.12	2.21	73.2-84.4

Note: Variables defined in Tables 3 and 4. Data are log differences of quarterly averages. OLS regressions run as a 3rd order polynomial distributed lag on right-hand side variables: 12 lagged observations with omission of concurrent observation. Regression coefficients above are the sum of the 12 estimated coefficients for each lag.

TABLE 3.6
World Money and U.S. Tradeable Goods Prices (WPI) Under Floating Exchange Rates: 1973.2 to 1984.4 (quarterly data, t-statistics are in parentheses)

$\dot{M}^{W}$	$\dot{M}^{ROW}$	$\dot{M}^{US}$	$\dot{E}^{US}$	$\bar{R}^2$	SER (Percentage points)	D.W.
3.11 (5.49)				0.45	1.12	1.80
	1.49 (4.73)			0.39	1.32	1.69
		0.81 (0.70)		-.04	1.73	0.98
			-0.84 (-6.02)	0.50	1.21	1.95
1.06 (0.77)			-0.62 (-1.86)	0.46	1.25	2.05
	0.03 (0.05)		-0.80 (-2.60)	0.46	1.25	2.05
		1.20 (1.35)	-1.07 (-5.17)	0.49	1.12	2.21
	1.83 (5.38)	2.38 (2.39)		0.42	1.29	2.01

Notes: Detailed definitions of variables are in Table 3. $\dot{WPI}$ is dependent variable: growth in the U.S. wholesale Price Index as defined by line 63 of IFS. $\dot{M}^{W}$ is percentage growth in world (narrow) money: 11 industrial countries. $\dot{M}^{ROW}$ is percentage money growth in 10 countries other than U.S. $\dot{M}^{US}$ is U.S. narrow money: M1. E is the IMF's in index of the dollar exchange rate: foreign currency/dollars "merm" weighted against 17 other industrial countries. Data are log differences of quarterly averages. OLS regressions are run as an unconstrained 3rd order polynomial distributed lag on the right-hand side variables: lagged 12 quarters excluding concurrent one. The regression coefficients above are the sum of the 12 estimated coefficients for each lag.

TABLE 3.7
World Money Variables and U.S. GNP Price Deflator Under Floating Exchange Rates: 1973.2 to 1984.4 (quarterly data, t-statistics in parentheses)

$\dot{M}^{W}$	$\dot{M}^{ROW}$	$\dot{M}^{US}$	$\dot{E}^{US}$	$\bar{R}^2$	SER (Percentage points)	D.W.
0.76				0.50	0.43	1.22
(3.87)						
	0.36			0.48	0.45	1.20
	(3.40)					
		0.44				
		(1.12)				
			-0.30	0.53	0.43	1.23
			(-6.15)			
-0.32			-0.24	0.66	0.36	1.70
(-0.79)			(-2.47)			
	-0.19		-0.30	0.64	0.37	1.67
	(-1.06)		(-3.27)			
		0.57	-0.34	0.55	0.41	1.33
		(1.91)	(-4.87)			
	0.44	0.62		0.52	0.43	1.39
	(3.92)	(1.87)				

Notes: Detailed Definitions of Variables are on Table 3. $D\dot{E}F^{US}$ is dependent variable: growth in U.S. GNP deflator $\dot{M}^{W}$ is percentage growth in world narrow money: 11 industrial countries. $\dot{M}^{ROW}$ is percentage money growth in 10 countries other than U.S. $\dot{M}^{US}$ is U.S. narrow money: M1. $\dot{E}^{US}$ is the IMF's index of the dollar exchange rate: foreign currency/dollars "merm" weighted against 17 other industrial countries. Data are log differences of quarterly averages. OLS regressions are run as an unconstrained 3rd order polynomial distributed lag on the right-hand side variables: lagged 12 quarters excluding concurrent one. The regression coefficient above are the sum of the 12 coefficients for each lag.

Table 3.5 shows the results of running equation (1) as a 12-quarter, third degree, polynomial distributed lag--and is based on 44 observations from 1962 to the first quarter of 1973 when the float began.[5] The concurrent observation on M is omitted to minimize simultaneity. These quarterly regressions tell much the same story: $\bar{R}^2$ is 0.47 for the WPI regression, and 0.61 for the GNP deflator. These equations are well behaved with no serial correlation in the residuals and the sum of the coefficients on the money supply are highly significant--and considerably greater than that shown in the annual regressions. The WPI reacted a bit faster to U.S. money growth than did the GNP deflator--as one would expect from the way in which the two indices are constructed.

In short, besides being the instrument by which monetary policy is conducted, U.S. M1 was itself a robust indicator of future cyclical price inflation in the United States. In the 1950s and 60s, when exchange rates were largely fixed, the system behaved as if the demand for American money were stable.

The Collapse of Domestic Monetarism in the 1970s and 80s

For the period of floating exchange rates from 1973 to 1984, however, consider running the same regressions fitting equation (1) for the United States. Now the good statistical fit for P on M completely disappears!

In comparison to the earlier period, Tables 3.3, 3.4 and 3.5 show the sharp reductions in $\bar{R}^2$ which in most cases becomes insignificant; and the signs of the 'a' coefficients are now often negative in Tables 3.3 and 3.4 based on annual data; and serial correlation in the 'u' residuals is much more marked--particularly in Table 3.5 based on quarterly data.

Apparently, one can no longer predict the now much-larger cyclical fluctuations in the U.S. price level by looking at changes in U.S. money growth by itself.

Several hypotheses might explain this breakdown in the domestic monetarist equation in the 1970s and 1980s--including the more rapid pace of domestic financial innovation causing M1's velocity to shift, oil shocks, and so on. Let us, however, proceed to test the proposition that shifts in international portfolio preferences destabilize (d) the demand for money in the United States (or at least reflected any shifts that did occur), and that the dollar exchange rate is a useful leading indicator of such changes.

The Dollar Exchange Rate

Consider amending our basic regression equation to incorporate the dollar exchange rate as an additional explanatory variable.

$$\dot{P} = C + \sum_{i=0}^{n} a_{-i}\dot{M}^{US}_{-i} + \sum_{i=0}^{n} b_{-i}\dot{E}^{US}_{-i} + v \qquad (2)$$

E is the International Monetary Fund's measure of the dollar exchange rate trade ("merm") weighted against 17 other industrial countries. Because E is measured in foreign currency units per dollar, $\dot{E}$ being positive represents dollar appreciation. Thus one would expect the 'b' coefficients in equation (2) to be negative. An appreciation of the dollar portends future reductions in U.S. price inflation because:

(i) The effective demand for U.S. dollar assets in general and U.S. money in particular has increased; and

(ii) Foreign goods will now be cheaper in dollar terms, putting downward pressure on American tradeable goods prices; and

(iii) Money growth in other industrial countries tends to decline--adding to the worldwide deflationary pressure.

Obviously, our simple regression equation (2) cannot distinguish among these three interrelated effects. But neither need the Federal Reserve in order to stabilize better the U.S. price level by making use of the information contained in the exchange rate.

For the period of floating exchange rates from 1973 to 1984, $\dot{E}$ turns out to be highly significant as shown in Table 3.3, 3.4, and 3.5. The 'b' coefficients are significantly negative and $\bar{R}^2$ is high and positive when equation (2) is run on annual or on quarterly data, and when either the WPI or the GNP deflator are the dependent variable. Indeed, the robustness of the dollar exchange rate as a leading indicator of future American price inflation is quite remarkable.

Focusing first on Table 3.5 based on quarterly data (with a 12-quarter polynomial distributed lag on both $\dot{E}$ and $\dot{M}$), one can see that the sum of the 'b' coefficients is -0.34 for the GNP deflator, and -1.07 for the WPI. A one percent increase in the dollar exchange rate will eventually reduce price inflation in U.S. tradeable goods (the WPI) by about one percentage point, and reduce inflation in the U.S. GNP deflator by about one third of that. These are large numbers because it is not unusual for the dollar exchange rate to move as much as 10 or 20 percent in a year. Indeed, the "effect" of shifts in the dollar exchange rate on cyclical changes in

American price inflation seems much larger than that which can be explained solely by international commodity arbitrage or foreign money growth under points (ii) and (iii) above. A positive E could also indicate deflationary pressure <u>within</u> the United States--as if the demand for U.S. money were changing.[6]

Tables 3.3 and 3.4, based on annual data, show that the effect of E on the WPI is somewhat more immediate--taking place early in the second year after the dollar exchange rate changes and continuing into the third. Whereas, E's impact on the GNP deflator is stretched out more toward the end of the second and into the third year.

Figure 5 shows the negative impact of $\dot{E}$ on changes in the U.S. WPI after 5 quarters. The simple correlation coefficient between the unsmoothed $W\dot{P}I$ and $\dot{E}$ (lagged) is -0.53; whereas when both series are smoothed this negative correlation increases to an astonishing -0.82.

Finally, the incorporation of the exchange rate into our basic regression equation run for 1973-84 makes the 'a' coefficients associated with U.S. M1 more sensible: they become positive and closer to being statistically significant. Indeed, the (spurious) negative coefficients for the first 4 or 6 quarters after the money supply changes (without the exchange rate in the equation) take on normal positive signs. This improvement is likely associated with the reduction of serial correlation in the residuals once the exchange rate is introduced. Serial correlation often reflects the influence of an omitted independent variable.

In summary, from the 50s and 60s to the 70s and 80s, the great deterioration in the quality of our basic monetary equation for the United States is avoided once the dollar exchange rate is included as an additional explanatory variable.

A CAUTIONARY NOTE

Because of the inherent asymmetry in the world dollar standard, monetary equations like (1) or (2) above need not fit at all well for countries other than the United States--such as any European country or Japan.

First, as we have seen, other countries domestic money growth rates are much more endogenized to the state of the foreign exchanges. Thus M is not truly an independent right-hand side variable.

FIGURE 5
U.S. EFFECTIVE EXCHANGE RATE AND WPI

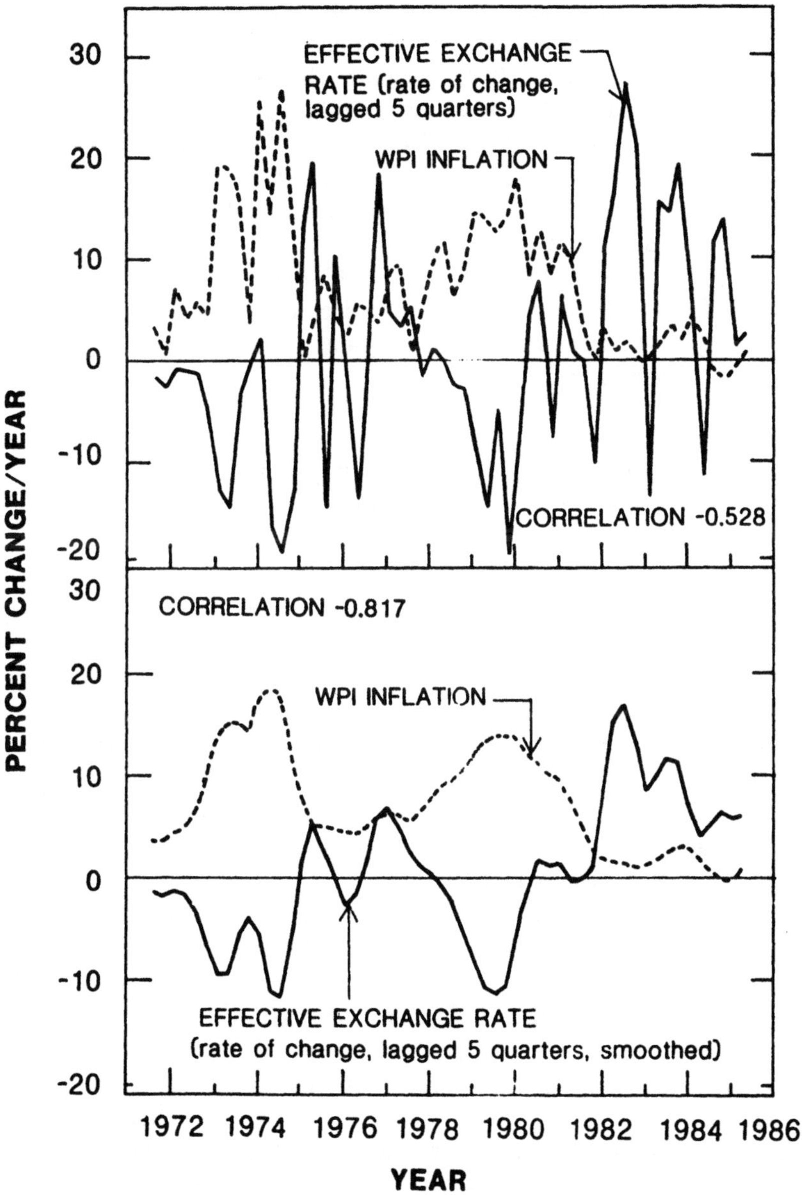

Secondly, when other countries exchange rates are strong (and the dollar is weak) these are also times of international inflationary pressure emanating from the United States throughout the world economy. Thus, the domestic deflationary pressure from an appreciating (non-American) currency is obscured.[7]

For example, although the European currencies had sharply appreciated in the late 1970s against the U.S. dollar, they still suffered the worldwide inflation of 1979-80--albeit in a more muted fashion than the United States (Table 3.2). Thus the exchange rate could have the "wrong" sign if one applied a regression model such as equation (2) to, say, Germany because of the influence of the international business cycle.

Only for the center county, the United States, does equation (2) apply for the 1970s and 80s. America has the only "independent" monetary policy, and its exchange rate fluctuations governed the international business cycle.

WORLD MONETARY VARIABLES AND U.S. PRICE INFLATION UNDER FLOATING RATES

Because of the inverse correlation between the strength of the U.S. dollar and money growth in the rest of the world under "dirty" floating, the explanatory variable E in equation (2) already captures much of the impact of worldwide inflationary or deflationary pressure. But can the Fed obtain yet more useful information about the future American price level by looking directly at money growth in other industrial countries?

I have argued that changes in the demand for dollar assets in general, and for U.S. money in particular, are manifested in the foreign exchange market in two ways:

-under predominantly "clean" floating, by fluctuations in the average dollar exchange rate against other major currencies; and

-when other countries' central banks act to smooth their dollar exchange rates, by fluctuations in foreign money growth.

In the latter case, changes in growth of foreign "hard" moneys--which are to some extent substitutable for dollars in international asset portfolios (McKinnon, 1982)--may itself have an additional inflationary impact on internationally tradeable goods in the world at large. And indeed, Table 3.2 shows the remarkable positive correlation in cycles of price inflation across the industrial countries. This then feeds back on the U.S. price level.

So foreign money growth, under the world dollar standard, both reflects changing money demand in the United States and has its own supply side effect on the world price level. And the simple regression models presented below cannot pretend to disentangle these two effects.

Tables 3.3, 3.4, 3.6 and 3.7 present the results of running regression equations of the form:

$$\dot{P}^{US} = C + \sum_{i=0}^{n} \dot{M}^{W}_{-1} + u$$

$$\text{or } \dot{P}^{US} = C + \sum_{i=0}^{n} \dot{M}^{ROW}_{-1} + v \qquad (3)$$

$\dot{M}^{W}$ is percentage growth in the "world" money, including U.S. M1 with a heavy weight, as shown in Table 3.1 for annual data; and $\dot{M}^{ROW}$ is money growth in the 10 industrial countries other than the United States portrayed in Table 1.[8]

In the 1970s and 1980s, world money does much better than U.S. money in predicting either the U.S. WPI or the GNP deflator: the regression coefficients for $\dot{M}^{W}$ are highly significant. The effect of world money on American tradeable goods prices (Table 3.6) is greater than its effect on the American GNP deflator (Table 3.7) as one would expect.

Even ROW money by itself does considerably better than U.S. money by itself in predicting U.S. prices as--Tables 3.6 and 3.7 based on quarterly data make clear. Moreover, the explanatory power of U.S. money improves substantially when ROW money is included as an additional explanatory variable (tables 3.6 and 3.7)--as if it were indeed proxying for shifts in the domestic demand for American M1.

In summary, money growth in the rest of the world does seem to be important, and there is a prima facie case for the Fed to take other countries monetary policies into account when formulating its own.

Under present world monetary arrangements, however, the dollar exchange rate seems to dominate these world and ROW money supply variables. Suppose E is added as an additional explanatory variable, and regressions are run in the format:

$$\dot{P}^{US} = C + \sum_{i=0}^{n} \dot{M}^{W}_{-i} + \sum_{i=0}^{n} \dot{E}^{US}_{-i} + u$$

$$\text{or } \dot{P}^{US} = C + \sum_{i=0}^{n} \dot{M}^{ROW} + \sum_{i=0}^{n} \dot{E}^{US}_{-i} + v \qquad (4)$$

Then, Tables 3.6 and 3.7 show that the E variables remain significant with (correct) negative signs, but M^W and M^{ROW} become insignificant with sometimes the wrong signs. This dominance of the dollar exchange rate is undoubtedly related to its inverse correlation with the world money variables.

As a first approximation, therefore, the Fed could treat the dollar exchange rate by itself as its primary signal of when American monetary policy was too tight or too easy provided that the reactions of foreign central banks remain similar to what they have been in the past.

IMPLICIT VERSUS EXPLICIT MONETARY COORDINATION WITH OTHER COUNTRIES: A CONCLUDING NOTE

Clearly, the U.S. Federal Reserve System should take a more open-economy approach to the problem of stabilizing the U.S. price level. But it would be a mistake to completely jettison monetarist rules governing domestic money growth: people still need forward assurance of what the monetary authority plans to do. A more ad hoc monetary strategy, even one where the dollar exchange rate was given some (indeterminate) weight, could add to uncertainty about the future and make the current demand for dollar assets--including money--more volatile.

Consider the following simple rules which could be unilaterally announced by the American authorities:

(1) The Fed would continue for the year ahead to project "normal" noninflationary growth in the major U.S. monetary aggregates--say 4 to 6 percent growth in M1.

(2) However, if the dollar was unusually strong in the foreign exchange markets, U.S. money growth would increase beyond its norm until the dollar came down--and vice versa.

If it had followed such a procedure, the Fed could have greatly ameliorated--perhaps largely avoided--the two great inflations of

1973-74 and 1979-80 by contracting in 1971-72 and again in 1978-79. Similarly, by expanding more in late 1981 and early 82, the Fed could have avoided the unusually rapid deflation of 1982-83.

Most recently, by failing to respond to the sharp run-up of the dollar in 1984 by monetary ease, the Fed imposed undue deflation on American tradeable goods industries and a slowdown in real growth in the American economy in 1985. The Fed eased in 1985, but that was a bit late given that exchange rate signal occurred much earlier.

Under (2) above, the Fed could go one step further. Exchange rate targets against hard foreign moneys could be made more precise through some purchasing power parity calculation. Elsewhere, I and others (McKinnon, 1984 and Williamson, 1983) have suggested "soft" target zones--for example, aiming to keep the dollar within 2.1 to 2.3 marks, and between 200 and 220 yen in 1985.

Once the dollar moved outside these zones, the Fed would be obligated to alter its monetary stance. If the Fed clearly announced its new strategy, private expectations would then more readily coalesce around what the exchange rate was likely to be--making it naturally more stable. Protectionist pressure in the American economy would abate once the 'real' price of dollars in terms of foreign currencies was confined to a narrow band which properly aligned the American price level with those prevailing in other industrial countries.

Although I believe that having the Fed unilaterally key on the dollar exchange rate would better stabilize the U.S. price level (and the world economy more generally), this hypothesis does rest on the assumption that implicit monetary cooperation by other central banks will continue. That is, when the dollar is unusually strong, other industrial countries would slow their money growth to smooth their exchange rate--and then speed up when the dollar became weak--as Figure 2 indicates they have done in the past.

However, suppose now the Fed officially adopts our new monetary strategy of keying on the dollar exchange rate without any explicit agreement on international monetary coordination. Although not necessarily likely, other central banks might now relax and not take symmetrical action to smooth their dollar exchange rates. Let the Fed do it!

For example, if in 1984 the Fed had embarked on a major monetary expansion in response to the strong dollar, other central banks might have expanded in parallel--or at least not contracted as they actually did (Figure 2). Then, not only would the dollar not have come down in the foreign exchange market, but there could have been too much monetary expansion overall--leading to worldwide inflation in 1985-86.

To deal with this dilemma, the Fed could informally monitor what other central banks are doing. If they (unexpectedly) expanded in parallel with the Fed when the dollar was strong, the Fed would be forced to lay off somewhat and give the exchange rate less weight.

Far better to secure an explicit agreement among the Fed, the Bank of Japan, and the Bundesbank (representing the European bloc) to react symmetrically to pressure on the dollar exchange rate.[9] Under such an agreement, only the Fed would be forced to revise substantially its operating procedures from an 'insular' to an open-economy mode. And, international altruism aside, having the Fed key on the dollar exchange rate would be very much in America's own best interests.

NOTES

I would particularly like to thank Ken Ohno for his unflagging computer and conceptual support--without which this paper could not have been written in its present form.

1. Proposals for the United States to return to a gold standard are not easy to classify.

If the U.S. unilaterally adopts a gold standard at some fixed parity, this fails to guarantee stability of the U.S. price level in terms of a broader commodity price index or guarantee stability of exchange rates with major trading partners. The price of foreign goods could still fluctuate widely in dollar terms. Indeed, without (symmetrical) monetary adjustments by other countries, an "equilibrium" gold parity would be difficult for the U.S. by itself to maintain.

On the other hand, if all the major countries agree to go back to gold simultaneously, then proper monetary coordination among them could, conceivably, maintain fixed exchange rates and common price level. However, international monetary coordination could be achieved without a gold cover--see the last section of this paper and McKinnon (1984).

Even if gold parities could be mutually established across several countries, a gold standard system would leave no discretionary mechanism for dealing with worldwide inflation or deflation--depending on what the monetary demand for gold turned out to be. And the demand for gold would be particularly unstable during the transition period when major countries were deciding on whether or not to reestablish gold convertibility.

2. Unless all countries are "small" with completely independent monetary and financial policies. Then the statistical law of large numbers would smooth out all foreign portfolio disturbances in any one country's average exchange rate with the outside world. Those exchange fluctuations that remained would then be uniquely associated with domestic financial disturbances within the country in question. But this extreme form of symmetry is hardly consistent with the "large" American economy's position at the center of the world dollar standard as described below.

3. In general, complete price-level and exchange-rate stabilization across the hard-currency industrial countries requires full scale monetary coordination. Either the Fed or the Bundesbank, or perhaps both, should adjust their national money growth rates in response to pressure on the mark/dollar rate. And elsewhere I have spelled out (McKinnon, 1974 and 1984) how such a first-best monetary agreement among Germany (representing the European bloc), Japan, and the United States could work.

In the test, however, we are considering a more limited, "second-best," approach. Suppose reactions of other central banks to exchange rate changes continue more or less as they have since floating began in early 1973. Is then the average dollar exchange rate a potentially useful monetary indicator for the Fed by itself?

4. Canada is the major exception. Because its currency is more closely tied to the American, the Canadian dollar does not provide international speculators with much of a portfolio alternative to holding U.S. dollars.

5. The results of estimating these equations and the subsequent ones are invariant to the choice of the distributed lag--whether 3rd or 4th degree polynomial, with or without end point constraints, and so on (Ambler, 1985).

6. A full theoretical description of how the effective demand for domestic money might change in response to portfolio shifts in the international bond market--the principle of indirect currency substitution--is provided in McKinnon (1985).

7. This inherent asymmetry between the United States and other countries was not understood by Franco Spinelli (1983) in his strong criticism of my open-economy approach to monetary stabilization. Moreover, Spinelli defined his "World" monetary

variables incorrectly. For a more complete analysis and rebuttal of Spinelli's work, see Bulchandani and Ohno (1985).

8. I have used fixed (mid period) GNP weights--and not fluctuating exchange rates--for constructing these world money aggregates. This permits us to distinguish the exchange-rate from the world-money supply in our regressions explaining the U.S. price level. Apart from this statistical convenience, however, there is a strong economic rationale for focussing on this definition of world money (Table 3.1) as a potential control variable for the world price level--see McKinnon, 1984, Ch. 5.

9. In Chapter 5 of An International Standard for Monetary Stabilization (1984), I have outlined a more complete set of rules as one possible basis for such an agreement. The ultimate objective is to secure the mark/dollar and yen/dollar exchange rates, while stabilizing the three countries' common price level measured in terms of tradeable goods.

REFERENCES

Ambler, S. "Comment on International Factors and U.S. Inflation: An Empirical Test of the Currency Substitution Hypothesis." (unpublished) Stanford University, May 1985.

Bulchandani, R. and K. Ohno. "World Money, the Exchange Rate and Tradeable Goods Prices: Some International Evidence." (unpublished) Stanford University, May 1985.

Dornbusch, R. "Expectations and Exchange Rate Dynamics", Journal of Political Economy. 84, 6(December 1976): 1161-1176.

Frenkel, J. and M. Mussa, "Asset Markets, Exchange Rates, and the Balance of Payments," Handbook of International Economics. R. Jones and P. Kenen (eds.) Vol.2 North Holland, Amsterdam, 1985.

Frenkel, J. "Purchasing Power Parity: Doctrinal Perspective and Evidence from the 1920s," Journal of International Economics, 8(May 1978): 169-191.

Friedman, M., "Commodity-Reserve Currency", Journal of Political Economy, 59(June 1951): 203-32.

_______. A Program for Monetary Stability. Fordham University Press, 1960.

_______. "The Role of Monetary Policy" American Economic Review. March 1968. Reprinted as Ch. 5 in M. Friedman The Optimum Quantity of Money and Other Essays. Aldine Publishing Co., Chicago, 1969.

Graham, F. D. Social Goals and Economic Institutions. Princeton: Princeton University Press, 1942.

Hall, R. "Optimal Fiduciary Monetary Systems", Journal of Monetary Economics. 12(July 1983): 33-50.

Hart, A. G. "The Case as of 1976 for International Commodity-Reserve Currency", Weltwirtschaftliches Archiv. Band 112, (1976).

Kenen, P., "The Role of the Dollar as an International Currency" Occasional Papers. No. 13. New York: Group of Thirty, 1983.

Levich, R. "Empirical Studies of Exchange Rates: Price Behavior, Rate Determination and Market Efficiency". Handbook of International Economics, R. Jones and P. Kenen (Eds.) Vol. 2, North Holland Amsterdam, 1985.

McKinnon, R. I. "A New Tripartite Monetary Agreement or a Limping Dollar Standard?" Princeton Essays in International Finance. No. 106, Princeton, New Jersey, 1974.

_______. Money in International Exchange: The Convertible Currency System. Oxford University Press, New York, 1979.

McKinnon, R. I. "The Exchange Rate and Macroeconomic Policy", Journal of Economic Literature. 72, 30(June 1981):531-557.

________. "Currency Substitution and Instability in the World Dollar Standard", American Economic Review. 72, 80(June 1982): 320-333.

________. An International Standard for Monetary Stabilization, Institute for International Economics, Washington, D.C., 1984.

________. "Two Concepts of International Currency Substitution", The Economics of the Caribbean Basin. M. Connolly and J. McDermott (Eds.), Praeger, New York, 1985.

Meese, R. A. and K. Rogoff. "Empirical Exchange Rate Models of the 1970s: Do They Fit Out of Sample?", Journal of International Economics. 14(February, 1983):1-24.

Ohno, K. "Purchasing Power Parity in a Financially Integrated World: A Re-Examination of Causality", (Unpublished) Stanford University, May 1985.

Spinelli, F. "Currency Substitution, Flexible Exchange Rates, and the Case for Monetary Cooperation: Discussion of a Recent Proposal", International Monetary Fund, Staff Papers. Vol. 30, No. 4, December 1983.

Williamson, J. The Exchange Rate System. Institute for International Economics, Washington, D.C., 1983.

PART II

MACROECONOMIC - AGRICULTURAL LINKAGES IN TRADITIONAL FRAMEWORKS

4

Inflation and Agriculture: a Monetarist-Structuralist Synthesis

Shun-Yi Shei and Robert L. Thompson

In the past agricultural economists devoted considerable attention to analyzing and attempting to understand the relationship between inflation[1] and agriculture. After being viewed as a moderator of inflation in the U.S. economy for more than two decades, the early 1970s found agricultural prices suddenly exploding upwards, leading the upward spiral of the general price level. Agricultural economists' attention was drawn to attempting to account for the unexplained large increase in the price of agricultural commodities relative to other prices in the economy. As the decade wore on, agricultural prices retreated from the heights attained in 1973, however the general price level continued to increase at significantly higher than recent normal rates.[2] A number of explanations of the large commodity price increases have been offered because the usual tools of price forecasting and analysis failed badly when it came to predicting nominal price behavior in the decade of the 1970s.

Many of the explanations offered to account for the events of 1973 have been based upon structuralist arguments (e.g., Hathaway, Schnittker). The most common explanation is as follows. Adverse weather conditions in many parts of the world, including the USSR, reduced agricultural production significantly below trend. Demand for agricultural products was growing at a rapid rate due to rapid population and economic growth. The industrialized economies were

Shun-Yi Shei, Associate Research Fellow, Institute of Economics, Academia Sinica, Nankang, Taipei, Taiwan. Robert L. Thompson, Dean of Agriculture, Purdue University, W. Lafayette, IN.

in a lockstep economic boom. The Soviet Union changed its policy and elected to import large quantities of grain rather than slaughtering livestock. Given the short-run inelasticity of supply and demand, the downward shift in supply caused a veritable price explosion in the face of growing demand. This effect was reinforced by import policies in a number of countries which cut the link between domestic and world market prices. As a result, internal prices in many countries did not rise at all to help ration the smaller supply in that short crop year.

This agricultural price increase is viewed as having been quickly transmitted through the economy to produce a still larger rate of inflation because both the labor market and industrial sector are considered to be less than perfectly competitive. The increase in consumer prices associated with rising food prices gets translated into higher wage rates via escalator clauses in wage contracts. The industrial sector, which is also viewed as imperfectly competitive, is assumed to employ "cost-plus" pricing so that increased costs, whether due to higher wage rates or higher input costs, get quickly passed through in the form of higher output prices. This raises agricultural input costs and so the spiral continues. It is assumed that the monetary authority, i.e. the Federal Reserve System, passively accommodates the increases in the general price level that originated, in this case, in the agricultural supply shock (Olivera).[3]

The monetarist approach, which argues that inflation is caused by autonomous, rather than accommodating, increases in the money supply, has been relatively little heard as an explanation for the agricultural price increases of the 1970s. Two exceptions to this generalization, are van Stolk and Houthakker, who asserted that the rapid growth by historical standards in international liquidity starting in the late 1960s made a significant contribution to raising agricultural prices in the early 1970s.

Schuh presented a related argument in which the linkage between money supply and agricultural prices is indirect via the exchange rate. He argues that excessive expansion in the U.S. money supply beginning in the 1960s, particularly associated with the "guns and butter" policy during the Vietnam War, culminated in the devaluations of the U.S. dollar in August 1971 and February 1973. This in turn lowered the price of our agricultural exports to foreign purchasers. From the U.S. perspective this had the effect of shifting the export demand schedule which we confront upwards and in turn raising domestic market prices. Chambers and Just have specified and estimated an econometric model, the results of which support Schuh's hypothesis. Soe Lin provides similar evidence in the Canadian case.

It is somewhat surprising that the direct interpretation of the monetarist approach has not been employed more often as an explanation of commodity price increases in the 1970s in light of the intense activity in that decade associated with the "monetarist counterrevolution" (e.g., Johnson) and monetary approach to the balance of payment (e.g., Frenkel and Johnson), together with the increasing evidence from Granger causality tests (e.g., Sims, Khan) of changes in the money supply leading changes in the general price level.

Perhaps the most likely explanation for the failure to consider the monetary approach more in accounting for sectoral price changes has been the conventional assumption of the final neutrality of money, that is, that increases in the money supply raise all prices by the same proportion, leaving relative prices unchanged. In light of this assumption, other explanations were sought to account for changes in relative prices, such as the large increase in agricultural commodity prices in 1973 relative to other prices in the economy. However, in the past several years theoretical arguments have been advanced that monetary shocks have non-neutral effects on relative prices and that, in fact, the conditions required for neutrality to hold, at least in the short run, are very stringent indeed (Bordo, pp. 1089). These theoretical arguments all assume that some prices in the economy are more flexible than others within any given period of time. That is, adjustment speeds differ among sectors. This is attributed variously to structuralist assumptions about differences in the competitive structure of the various industries (Sayad) or to the existence of implicit contracts of varying duration (Bordo).[4] They have demonstrated that under such conditions, a monetary shock will have real effects, i.e. cause relative prices to change. Bordo has carried out regression tests which demonstrate significant differences in adjustment speed among sectors. Barnett, Bessler, and Thompson found significant causal relationships between both domestic money supply and international reserves and nominal agricultural prices (1981a) and the food relative to the nonfood components of the U.S. CPI (1981b).

It should be clear that these latter efforts represent in a sense a rapprochement between the monetarist and structuralist approaches to accounting for inflation. They recognize that there exist structural differences among sectors of the economy which affect the transmission of inflationary shocks, regardless of where they originate. On the other hand, they acknowledge that changes in the money supply may be an autonomous causal force and not only a passive accommodator of price increases which originated in the real sector. It should be borne in mind that tests of causality such as those cited above do not imply that changes in the money supply are the only cause of the increase in prices, but rather that there exists a significant causal flow from money to prices. There is no known study in the literature which attempts to estimate the relative importance of the respective explanations that have been offered, in particular: the world-wide crop shortfall, the Soviet grain purchase, the U.S. dollar devaluation, and the expansion in the U.S. money supply.[5]

The first objective of this paper is to clarify the linkages through which inflation gets transmitted among the real sectors of the economy and between them and the monetary sector. In the process, a synthesis between the monetarist and structuralist approaches is attempted. It is necessary to do this in somewhat more detail than in the theoretical papers of Sayad and Bordo in order to specify a formal econometric model of the U.S. economy in which the linkages are explicit. The second objective is to illustrate the utility of this approach by applying it to assess the relative roles of the above four variables which have been hypothesized to "cause" the observed large price increases in 1973.

The approach taken here is to specify a four-sector (crops, livestock, industry, services) open general equilibrium model of the U.S. economy which also includes a simultaneous monetary sector. The real sector specification is consistent with the structuralist approach, in that agricultural prices are assumed to be competitively determined, but industrial and service sector prices are less flexible. The monetary approach, and in particular the absorption variant (Alexander), is then drawn upon to specify the monetary sector and its linkages to the real sectors.

Once specified and estimated, the econometric model is employed to estimate the relative roles of the four variables identified above as contributing to the 1973 price increases. The static simulation results suggest that the largest single factor accounting for the rapid 1973 agricultural and general price inflation in the United States was the 10 percent expansion in the domestic component of the monetary base in that year. The results also provide evidence that dollar devaluation accounted for a significant part of the increase in U.S. prices, particularly in agriculture. The large Soviet grain purchase had the largest effect of the four explanatory variables on domestic crops prices, but relatively less effect on prices in the other three sectors. The results in general suggest that there exist significant simultaneities among sectors of the U.S. economy which affect the nature and speed of transmission of inflationary shocks. Monetary policy appears to have been an important omitted variable in accounting for agricultural price behavior in the 1970s.

THEORETICAL MODEL

This section develops a theoretical model which includes the principal linkages between agriculture and the rest of the economy. The model presented has both monetarist and structuralist characteristics. The model first recognizes that there exist structural rigidities in the economy and that prices do not adjust at the same speed in all sectors. Yet it also takes into account the fact that recent causality tests have provided evidence that monetary changes are not solely passive; that is only accommodating price increases associated with real shocks or imperfectly competitive markets. It recognizes that the general price level is a weighted average of the prices in the respective sectors of the economy, all of which are simultaneously determined. The focus here is on clarifying the linkages among the real sectors and between them and the monetary sector through which inflation is transmitted.

The Real Sector

For purposes of this study, the output of the real sector is divided into four aggregates: crops, livestock, industry, and services. The outputs from the first three sectors are tradeable, while services are assumed to be nontradeable. This distinction is important since in an open economy, the effects of a shock differ between internationally tradeable and nontradeable goods in both degree and speed of adjustment.[6] Within the set of tradeable goods industrial products are distinguished from agricultural products on structuralist grounds. Agriculture is viewed in the structuralist conception as a perfectly competitive sector which produces homogeneous goods whose prices are flexible to freely rise or fall as market forces dictate. Industry is viewed as being oligopolistic, producing heterogeneous goods under increasing returns to scale. Financial barriers impede entry of new firms. Prices are set on the basis of a profit margin over variable costs of production (i.e., "cost-plus pricing"). Nominal industrial prices, being cost-determined, tend to be inflexible downwards. The theory of implicit contracts, as applied by Bordo in this context, also provides a basis for this without having to assume imperfect competition.

Agricultural output is dichotomized into crop and livestock products for several reasons. This is necessary due to the basic difference in the supply response of the two. Supply response in livestock involves a capital stock adjustment decision and, in crops, an annual production input allocation decision. There is assumed to be no short-run supply response in the crop sector, due to the biological lag in production. While it is possible to increase slaughter of livestock in the short-run, the expected result from an upward shift in demand which raises expected prices is for farmers to retain animals from slaughter in order to expand the herd size to increase slaughter in the future. The short-run livestock supply response may therefore be negative. As a result, the price increases in the short-run in both cases are expected to be greater than the long-run.

The U.S. is a net exporter of crops and a net importer of livestock products. The domestic crops sector is quite open and exposed to world market forces, while the livestock sector is protected by trade barriers, particularly in the form of import quotas.

The model is general equilibrium in the sense that the equilibrium prices and quantities in the four sectors of the economy are simultaneously determined.[7] The general price level is a weighted average of the endogenously determined prices in the four sectors. Total national income is the sum of the price times output in the four sectors. Each sector must satisfy the market clearing condition that production plus imports must equal domestic absorption (consumption plus investment) plus exports.

A large-country assumption is maintained for the two export sectors, crops and industry. That is, it is assumed that the United States' market shares are sufficiently large that its actions have a perceptible effect on world-market prices.[8] Product homogeneity is assumed between the U.S. and the rest of the world. The law of one price is therefore assumed to hold such that domestic prices equal world market prices times the rate of exchange for crops and industrial goods, respectively. The livestock import quota cuts the link between domestic and world market livestock prices. These prices are therefore determined in the domestic market, just as in the case of services. As long as the livestock import quota is binding, no excess supply schedule of the rest of the world is needed.

The supply equations for the four sectors in this model are consistent with the structuralist specification outlined above. This suggests that the nature of the response of supply to changes in price differs among sectors. That is, agriculture is viewed as the flexible price sector. Crop sector output is specified as a function of the lagged real prices of crops and livestock and the effect of government programs which modify the area of land in crop production. Livestock supply, which is a slaughter or disinvestment decision is specified as a function of the real current prices of livestock and crops together with the herd size (i.e., the capital stock) at the end of the previous period.

In contrast, industry and services are specified as fixed price sectors, in each of which the change in price from one period to the next is made the dependent variable. Following Baumol, since the principal component of services is labor, its "supply equation" is specified as a price of services equation in which the percent change in nominal price is made a function of the percentage change in the aggregate nominal wage rate. That is, it is assumed that the nominal price of services moves up with the nominal wage. Industry is also viewed here as practicing cost-plus pricing. But, following Gordon, in the industry supply equation the percentage change in its nominal price is made a function of both the percentage change in nominal wage rate and the percentage change in the productivity index of the sector. This represents technological progress, capital accumulation, and economies of scale. The larger the productivity increase, the smaller the increase in price is expected to be for a given increase in wage rate. Therefore, the quantities of output of both industry and the services sector are determined by their respective demand equations, given the pricing equations.

The demand side of the economy is also specified somewhat differently from that to which analysts who specify supply and demand models on the basis of microeconomic theory are accustomed. First, following the absorption approach (Alexander) to the balance of payments, the dependent variable in each of the "demand equations" is the real domestic absorption of the given sector's output. Absorption is defined as the sum of domestic consumption, investment and government purchases. As in conventional demand equations, this is specified as a function of the real price of that sector's output and the real price of substitutes. However, real aggregate national expenditure is substituted in place of real per capital income in each equation. As will be clarified in the next section, per capita real aggregate national expenditure is not necessarily equal to real per capita income. In the absorption approach the national budget constraint requires that the difference between domestic expenditure and national income be exactly equal to the value of imports minus exports. It is through aggregate expenditures that money enters the system. It is assumed that domestic credit creation directly increases expenditures in this model. For example, the increased money supply may be viewed as paying for the government expenditures that Congress is unwilling to increase taxes to pay for directly. This is clarified in greater detail in the next section. The coefficient of this variable is interpreted as the marginal propensity to absorb a given sector's output out of an increase in aggregate expenditures. Differences in this marginal propensity across sectors represent one of the key explanations for why monetary shocks have different effects on the respective sectors in this model.

To close the real sector of the model all that remains is to specify export demand schedules for the crops and industrial sectors. Exports of each sector are specified net of imports. It is assumed that traded industrial and crop sector products are homogeneous and perfectly substitutable for foreign-produced goods of the respective sectors. This assumption is more limiting in the case of industrial than agricultural goods. The foreign export demand equation confronted by the United States in each sector is specified as a function of the respective U.S. export price (the domestic price times the exchange rate), the foreign consumer price index and real income, and other foreign supply or demand shifters. Since services are by definition non-tradeable, no export demand schedule is specified. No import supply schedule is specified for livestock products because import quotas effectively insulate the domestic market from world market price signals. The balance of trade is defined as the total value of exports of crops and industrial goods in foreign currency times the exchange rate. This completes the specification of the real sector of the model. The manner in which shocks get transmitted through it is discussed below after the monetary sector is specified.

The Monetary Sector

As briefly explained above, aggregate expenditures in the economy are not necessarily equal to national income within any given period. Rather, total expenditures equal national income plus some fraction of the difference between the actual stock of money in the economy and the long-run desired stock that economic agents wish to hold. For example, if the actual stock of money exceeds demand, aggregate expenditures will exceed national income as people "dishoard". On the other hand, if demand for real balances exceeds the supply, people's expenditures will be less than current income as they attempt to "hoard", i.e. to rebuild their real balances. This monetarist approach is based upon Mundell (1968, Chap. 8) and Prais (1961). It is assumed that full equilibrium is not achieved within one period, but rather that only a fraction of the difference between actual and desired holdings of real balances is eliminated within any given period.

Since total national income and the general price level are determined in the real sector, as discussed above, the function of the monetary sector model is to explain the difference between real national income and real aggregate expenditures, i.e. "dishoarding" or "hoarding". So, our objective here is to build the simplest possible monetary component of the model which will capture the essence of the adjustment between the supply and demand for real money balances.

The supply of money, represented here by M2, equals the monetary base (the stock of high-powered money) times the monetary multiplier. (We assume that a stable relationship exists to explain the monetary multiplier, but that the multiplier is not necessarily constant.) The monetary base is comprised of two parts: the domestic component and international reserves.[9] The domestic component equals the net liabilities of the central bank held as member bank deposits at the central bank, vault cash by non-member banks, and currency by the non-bank public. International reserves are comprised of official holdings by the monetary authority of gold, foreign exchange, and special drawing rights (SDRs), as well as the country's reserve position at the International Monetary Fund (IMF). Any change in the monetary base, then, whether from the domestic component or international reserves, alters the money supply through the monetary multiplier.

It is assumed here that the domestic component of the monetary base is exogenous to the model, i.e. that it is under the control of the Federal Reserve system. Clearly the Federal Reserve Board reacts to changing market forces by altering the size of the domestic component of the monetary base. Gordon (1975), among others, has provided a politico-economic explanation of the Fed's behavior in terms of the supply and demand for inflation, the equilibrium between which determines the change in the domestic component. However, in light of the causality tests which suggest that monetary adjustment is, at least to some extent, autonomous, and the interest of keeping the present model manageable, this variable was treated as exogenous. Recall that it is assumed that new money enters the system through government spending in excess of tax collections. Operationally, in the model it enters through autonomous increases in the money supply which then have to be either absorbed or held as real balances.

The stock of international reserves, which forms the other component of the monetary base, is altered within each period by the balance of payments (official settlements balance). This equals the balance of trade, determined in the real sector as defined above, plus net short-run capital inflows. International capital flows represent portfolio adjustments as economic agents attempt to maintain equilibrium in their financial asset portfolio as market conditions change. Short-run capital flows are viewed as a function of relative interest rates and relative rates of inflation in the United States and the rest of the world. (This is a very simplified conception. For a more detailed discussion and review of the literature, see Bryant.)

It is important to note that an increase in the official settlements balance of payments, whether from net capital inflows or an excess of export revenue over import expenditure, expands the monetary base and in turn the money supply unless the monetary authority offsets this change by autonomously altering the domestic component. (This process is known as sterilization.) This represents one of the critical links between the open economy and the rest of the world which reduce the freedom of action of individual governments in economic policy making. At a given rate of interest and expected rate of inflation, as real income increases, the transactions demand for real balances is expected to increase. However, at a given level of real income, as the opportunity cost of holding real balances, as reflected in the nominal interest rate rises, desired demand is expected to fall. Similarly, an increase in the expected rate of inflation, which would erode the value of existing real balances, would also cause a shift out of money into the consumption of real goods.

The public's desired demand for real balances is specified as a function of the real national income and the rate of inflation, both of which are determined in the real sector of the model, and the nominal rate of interest.

The model is specified under the assumption of fixed exchange rates. That is, it is assumed that the monetary authority stands ready to buy or sell any quantity of foreign exchange at the pegged price. It is through this channel that a positive balance of payments, for example, causes an increase in the domestic money supply. Under a freely floating exchange rate the balance of payments on official transactions is by definition zero and there can be no change in the stock of international reserves. The price of foreign currency, i.e. the exchange rate, freely moves up or down in response to the supply and demand (on both current and capital account) for foreign exchange. In this sense, a freely floating exchange rate in principle gives a government greater freedom of action in setting domestic economic policy. In reality, however, even though we abandoned fixed exchange rates at the end of 1973, the rates were never permitted in the 1970s to float freely without intervention from central banks' buying or selling on government account to modify the adjustments which market forces dictate. Moreover, one of the objectives of this study is to simulate the effects of dollar devaluation on prices in 1973. This requires that we be able to vary the exchange rate parametrically. It is judged that even in 1973 it was not particularly unrealistic to assume fixed exchange rates.

As indicated above, it is assumed that full equilibrium between supply and desired demand for real balances is not attained within one period. For example, if, as a result of an increase in the money supply from whatever source, the public's desired demand for real balances is less than the existing stock, the excess supply sets in motion an adjustment process through which expenditures increase. This starts to push up the general price level, thereby reducing the real value of the existing stock of nominal balances. Nevertheless, it is assumed that full adjustment is not attained within one period, yet the total stock has to be held at any point in time. For this to occur, the rate of interest must fall to induce the public to hold the existing stock.

In order to close the model, one must specify equations to explain the interest rate and the wage rate. For simplicity and to keep the model manageable, a decision was made not to specify a financial market component to explain the interest rate or a labor market component to explain the wage rate. Rather, quasi-reduced forms (which include simultaneously determined variables on the right hand side) are specified to explain these two variables. The interest rate is specified, following the previous paragraph, as a function of the variables which determine the excess supply (or demand) of real balances, real national income, the existing stock of real balances, and the rate of change in the general price level, and the lagged interest rate, assuming a disequilibrium specification is more appropriate than an equilibrium specification in these markets.

A Philips curve specification of the wage rate equation is made in line with the extensive research which has been carried out on the behavior of the wage rate in the United States. Following Wacher and Gordon (1975b), the percentage change in the wage rate between periods is specified as a function of the percentage change in the size of the monetary base and the unemployment rate. The rate of change in the stock of money reflects the upward pressure of excess demand in the real sector on the level of wages. The rate of unemployment reflects the downward pressure of excess supply in the labor market on the wage rate. Clearly, both equations are gross oversimplifications. However, it is impossible to model everything that would be desirable in detail and yet keep the model manageable. With these equations the monetary component of the model is complete, and the simultaneous solution of the relationships specified in the real and monetary sectors generate equilibrium values of all the endogenous variables in the system.

Transmission of Shocks

The four shocks cited in the introduction which have been offered as explanations for the agricultural price increases in the early 1970s fall basically in two categories. The worldwide crop failure, the Soviet grain purchase, and the dollar devaluation all shift the crop export demand schedule confronted by the United States upwards. The other explanation was the increase in the domestic component of the monetary base by the Fed. We will trace through the adjustments which each induces.

An autonomous increase in the money supply generates an excess supply over demand for real balances. In the model specified this sets in motion several types of adjustment. To induce the public to hold the existing stock of real balances, the interest rate must fall. However, this induces a capital outflow from the economy, reducing international reserves and thereby offsetting some of the original increase in the money supply.

Dishoarding by the public also occurs in the form of increased expenditures to reduce the stock of real balances held. This shifts the "demand" (absorption) equations upwards, putting upward pressure on crop and livestock prices. In the fixprice sectors, industry and services, the shifts in the "demand" schedules initially cause greater absorption to occur at the given price. Nevertheless, as soon as any nominal prices increase, the general price index rises, causing the wage rate to rise via the Philips curve relationship. This in turn increases industry and services prices. Simultaneously, the higher general price level lowers all real prices, further shifting all demand schedules and supply schedules. In light of the complex web of simultaneous interactions it is not possible to predict ex ante the net effect. The general price level is expected to increase in any case.

The higher domestic prices for exportable goods get transmitted through the assumed fixed exchange rate into higher prices in foreign currency to customers in the rest of the world. At the higher price they purchase a smaller quantity; the net effect on export revenue depends on the elasticity of export demand. If it is elastic this will lower export revenue. This is the more likely case.

It should be clear from this discussion that as a result of the monetary expansion, the balance of payments is likely to deteriorate on both capital and current accounts. If the exchange rate were permitted to float, it would depreciate under these circumstances, shifting the export demand schedules upwards and putting further upward pressure on the domestic prices of exportables.

An external shock such as an upward shift in the export demand for crops associated with a crop shortfall in the rest of the world, raises both the foreign and domestic nominal prices of crops and the balance of trade. The price of crops rises relative to the other sectors' prices, and the general price level rises. The higher general price level lowers the stock of real balances, which sets hoarding in process by the public in their attempt to rebuild the desired level of real balances. This also puts upward pressure on the interest rate, and in turn attracts capital inflows. This, together with the larger balance of trade, expands the monetary base setting in motion an adjustment process such as that described in the former case above. The empirical estimates are presented in the next section.

THE STATISTICAL RESULTS

Based upon the specification outlined above, a 24-equation econometric model was formulated. The model consists of 14 behavioral equations and 10 identities. The parameters of the simultaneous equations model were estimated by two-stage least squares using annual data from 1950 to 1974. All behavioral equations are linear in their parameters. The variables are defined and their data sources identified in Table 4.1.

The primary data sources were the USDA Agricultural Statistics, the U.S. Department of Commerce Business Statistics, and the IMF International Financial Statistics. For a detailed discussion of the operational definitions of all variables and of the data sources, and a listing of the data used in estimation, the reader is referred to Shei (pp. 141-152).

The empirical estimates of the parameters of each structural equation are reported in Tables 4.2 and 4.3 for the real and monetary sectors of the model, respectively. The t-values are presented in parentheses below their respective coefficients. Some experimentation was involved before the estimates presented here were obtained. In certain cases it was difficult to distinguish the relative effects of several explanatory variables due to a high degree of intercorrelation. As a result, several variables suggested by theory or logic were omitted from certain equations. The criteria for omitting variables were primarily the consistency of each parameter estimate's sign with a priori expectations and the size of the standard error relative to the parameter estimate.

The estimated coefficients in all of the domestic demand equations have expected signs with respect to their own real price and per capita real income variables. The cross price effect of livestock products in the domestic demand for crops equation has the expected positive sign. However, the cross price effect of crops in the domestic demand for livestock products had an unexpected negative sign and hence was dropped in this equation estimation. A trend variable was added to the demand for services equation because the data reveal a significant upward trend beyond that explainable by growth in per capita expenditures.

The price and income coefficients in the crop and industry export demand equations were unsatisfactory. The price elasticity of crop export demand was only -0.32, much lower in absolute value than estimates obtained in other studies and the income elasticity, 1.83, is much higher than values estimated in other studies. The estimated price elasticity of industrial export demand is -1.04; the estimate of the income variable had an unexpected negative sign. Table 4.4 summarizes several alternative estimates of export demand elasticities from different sources in the literature. Thompson (pp. 7-13) has summarized the problems associated with direct estimation of aggregate export demand equations as these. Due to the likely downward bias in the estimates derived, the two export demand

TABLE 4.1
Definition of Variables and Data Sources[a]

Notation	Operational Definition	Source
Endogenous Variables		
AE	Nominal national expenditure $= Y + XD^I + XD^C - MD^L$	
BOP	Official Settlements Balance of Payments $= XD^C - MD^L + XC^I + \Delta K + RIB$	
CPI	Consumer price index (calculated by using equation (12)), Table 2	
DD^C	Domestic absorption of crops $= DS^C - XD^C$	
DD^L	Domestic absorption of livestock products $= DS^L + MD^L$	
DD^I	Domestic absorption of industrial goods $= DS^I - XD^I$	
DD^S	Domestic absorption of services $= DS^S$	
DS^C	Cash receipts from total farm marketings of crops	USDA
DS^L	Cash receipts from total farm marketings of livestock and products	USDA
DS^I	Personal consumption expenditures on durable and nondurable goods less DS^C, DS^L, gross private domestic investment (fixed investment and change in business inventories), and XD^I	USDC
DS^S	Private and government purchase of services	USDS
HPM	Stock of high-powered money (= FR + NDA)	IFS
P^C	Wholesale price index of all grains	USDC
P^L	Wholesale price index of all livestock	USDC

TABLE 4.1 (continued)

Notation	Operational Definition	Source
P^I	Wholesale price index of all industrial goods	USDC
P^S	Consumer price index of services	USDC
PX^C	Export price index of crops	USDA
	= ag. export price index x value of ag. exports - / value of crop exports export price index of livestock x value of livestock exports / value of crop exports	Agricultural exports : Quantity index Foreign trade: Value of total agricultural exports and imports Ag. export price index = 100 x Ag. export value index / Ag. export quantity index
PX^I	Export price index of industrial goods = total commodity export price index x value of commodity exports / value of industrial exports ag. export price index x value of ag. exports / value of industrial exports	USDA
r	Nominal interest rate (Treasury bill rate)	IFS
W	Index of average hourly earnings (private nonfarm economy)	USDC
XD^C	Nominal value of U.S. net exports of crops: Calculated by aggregating the value of net exports of crops including cotton, nuts, fruits, grains, tobacco, oilseeds, vegetables	USDA--Foreign trade in agricultural products : values of exports and imports by principal groups
XD^I	Nominal value of U.S. non-agricultural commodity net exports (total value of non-agricultural exports--total value of non-agricultural imports)	USDA--Foreign trade in agricultural products: values of exports and imports by principal groups

TABLE 4.1 (continued)

Notation	Operational Definition	Source
RIB	Rest of items in U.S. balance of payments	IFS
T	Trend (1960=60, etc.)	
U	Total unemployment rate	USDC
Y_f	Gross national product in the world	UN
Z	International crop price linkage balancing factor = $\frac{PX^c}{e \cdot P^c} - 1$	
Z^I	International industrial price linkage balancing factor = $\frac{PX^I}{e \cdot P^I} - 1$	

[a]Unless specified otherwise, all unit values are in terms of billions of U.S. dollars and all indexes are based on 1967 = 100.

SOURCES:

IFS - International Monetary Fund, International Financial Statistics, 1971 Supplement and Nov. 1977, Washington, D. C.

USDC - United States Department of Commerce, Bureau of Economic Analysis, Business Statistics, Washington, D. C., May 1976.

USDA - United States Department of Agriculture, Agricultural Statistics, Washington, D. C., 1972 and 1975.

UN - United Nations, Statistical Yearbook (various issues), New York.

IMF - International Monetary Fund, Balance of Payments Yearbook (various issues), Washington, D.C.

USDL - United States Department of Labor, Bureau of Labor Statistics, Bulletin, 1926, Washington, D.C., 1977, p. 96.

FAO - Food and Agriculture Organization of the United Nations, Production Yearbook (various issues), Rome, Italy.

TABLE 4.2
Structural Estimates and Identities of the Real Sector

Domestic Absorption Equations

(1) $$\frac{DD^C}{PC} = \underset{(2.83)}{8.6381} - \underset{(2.36)}{0.0489}\frac{PC}{CPI} + \underset{(.61)}{0.0216}\frac{PL}{CPI} + \underset{(4.64)}{2.6738}\left(\frac{AE}{CPI}\right)/POP$$

(2) $$\frac{DD^L}{PL} = \underset{(10.38)}{15.0748} - \underset{(7.37)}{0.0798}\frac{PL}{CPI} + \underset{(21.58)}{4.4278}\left(\frac{AE}{CPI}\right)/POP$$

(3) $$\frac{DD^I}{PI} = \underset{(3.57)}{379.6889} - \underset{(5.44)}{4.4564}\frac{PI}{CPI} + \underset{(17.93)}{121.5216}\left(\frac{AE}{CPI}\right)/POP$$

(4) $$\frac{DD^S}{PS} = \underset{(5.79)}{498.1703} - \underset{(5.07)}{3.7856}\frac{PS}{CPI} + \underset{(4.98)}{36.5686}\left(\frac{AE}{CPI}\right)/POP + \underset{(9.47)}{11.7877}\,T$$

Export Demand Equations

(5) $$\frac{XD^C}{PX^C} = \underset{(1.43)}{-21.1967} - 0.0209\frac{PX^C}{CPI_f} + 0.025\left(\frac{Y_f}{CPI_f}\right) + \underset{(1.57)}{0.2563}\,POP_f$$

(6) $$\frac{XD^I}{PX^I} = \underset{(9.97)}{4.9231} - 0.0603\frac{PX^I}{CPI_f} + 0.0751\left(\frac{Y_f}{CPI_f}\right) - \underset{(10.21)}{0.7518}\,T$$

Domestic Supply Equations

(7) $$\frac{DS^C}{PC} = \underset{(38.50)}{13.0564} + 0.0234\frac{PC}{CPI} + \underset{(13.21)}{0.6678}\,T$$

(8) $$\frac{DS^L}{PL} = \underset{(30.13)}{13.0166} + 0.0751\frac{PL}{CPI} + \underset{(6.65)}{0.428}\,T$$

(9) $$\frac{\dot{PI}}{PI_{-1}} = \underset{(1.02)}{-3.0663} + \underset{(2.83)}{1.4445}\frac{\dot{W}}{W_{-1}} - \underset{(1.67)}{0.551}\frac{\dot{Q}}{Q_{-1}}$$

(10) $$\frac{\dot{PS}}{PS_{-1}} = \underset{(1.33)}{-0.9364} + \underset{(7.38)}{1.0195}\frac{\dot{W}}{W_{-1}}$$

Real National Income Definition

(11) $$\frac{Y}{CPI} = \frac{DD^C}{PC} + \frac{DD^L}{PL} + \frac{DD^I}{PI} + \frac{DD^S}{PS} + \frac{XD^C}{PX^C} - \frac{MD^L}{PM^L} + \frac{XD^I}{PX^I}$$

TABLE 4.3
Structural Estimates and Identities of the Monetary Sector

Net Capital Inflow Equation

(19) $\Delta K = \underset{(.99)}{-0.6895} + \underset{(2.74)}{2.9065}(r-r_{-1}) - \underset{(1.64)}{1.0231}(r-r_{-1})_f$

Official Settlements Balance of Payments Definition

(20) $BOP = XD^C - MD^L + XD^I + \Delta K + RIB$

Money Market Equilibrium Condition

(21) $HPM^S = FR_{-1} + NDA + BOP$

Aggregate Expenditures Equation

(22) $\frac{AE}{CPI} = \underset{(2.96)}{-21.5471} + \underset{(1.43)}{0.1924}\frac{HPM}{CPI} + \underset{(132.40)}{1.0006}\frac{Y}{CPI} + \underset{(1.47)}{1.0845}\,r - \underset{(3.16)}{0.7622}\frac{\dot{CPI}}{CPI_{-1}}$

Interest Rate Equation

(23) $r = \underset{(.63)}{-1.714} + \underset{(.68)}{0.1328}\,r_{-1} + \underset{(2.93)}{0.0067}\frac{Y}{CPI} + \underset{(1.39)}{0.122}\frac{\dot{CPI}}{CPI_{-1}} + \underset{(.03)}{0.0015}\frac{HPM}{CPI}$

Philips Curve Equation

(24) $\frac{\dot{W}}{W_{-1}} = \underset{(2.56)}{4.51} + \underset{(2.24)}{0.267}\frac{\dot{HPM}}{HPM_{-1}} - \underset{(.35)}{0.1167}U$

General Price Level Definition

(12) $CPI = \dfrac{P^C\frac{DS^C}{P^C} + P^L\frac{DS^L}{P^L} + P^I\frac{DS^I}{P^I} + P^S\frac{DS^S}{P^S}}{\frac{DS^C}{P^C} + \frac{DS^L}{P^L} + \frac{DS^I}{P^I} + \frac{DS^S}{P^S}}$

International Price Linkage Equations

(13) $PX^C = e \,.\, P^C\,(1 + Z^C)$

(14) $PX^I = e \,.\, P^I\,(1 + Z^I)$

Commodity Market Clearing Conditions

(15) $\frac{DS^C}{PC} = \frac{DD^C}{PC} + \frac{XD^C}{PXC}$

(16) $\frac{DS^L}{PL} = \frac{DD^L}{PL} - \frac{MD^L}{PML}$

(17) $\frac{DS^I}{PI} = \frac{DD^I}{PI} + \frac{XD^I}{PXI}$

(18) $\frac{DS^S}{PS} = \frac{DD^S}{PS}$

TABLE 4.4
Export Demand Elasticities for U.S. Crops and Industrial Goods, Alternative Estimates

Product Class	Price	Income
<u>Crops</u>		
Houthakker & Magee	-0.96	1.02
Clark	-0.38	n.a.
Hooper & Wilson	-0.88	n.a.
Hooper	-1.47	0.85
Tweeten	-6.4 to -16.0	n.a.
Johnson	-6.7	n.a.
<u>Industrial Goods</u>		
Houthakker & Magee	-1.22	1.17
<u>Finished Manufactures</u>		
Magee	-1.76	1.44

<u>Notes</u>: n.a. = not available

equations were reestimated with the price and income coefficients constrained to more plausible values chosen on the basis of the evidence presented in Table 4.4. The crop export demand equation was constrained to have a unitary price and income elasticity at the point of means. The price elasticity of export demand for industrial products was constrained to -1.5, and the income elasticity to 1.3.

The other area which proved most problematical in estimating the model was in the aggregate crop and livestock supply equations. As in previous econometric studies, it proved difficult to obtain satisfactory estimates of the price coefficients. Using annual data from 1911 to 1958, Griliches, with some difficulty, was able to conclude that the supply response of crops and livestock products to price is neither negative nor zero. He concluded that the price elasticity of aggregate crop supply was between 0.1 and 0.2, and of livestock supply, between 0.2 and 0.3. He also calculated the implicit aggregate farm supply elasticity on the basis of assumed elasticities of demand for inputs and obtained an estimate of 0.28 to 0.30 in the short run. Using acreage and yield response estimates, Tweeten and Quance calculated the U.S. aggregate supply elasticity for crops at 0.17 and for livestock at 0.38. The crop and livestock supply equations were reestimated with the price elasticities constrained to these values at the point of means.

The remaining equations in the real sector submodel presented no severe problems. The estimated coefficients in the industrial and services price equations had the expected positive signs.

Prior to estimation of the monetary model the demand for real balances equation was substituted into the aggregate expenditures equation to reduce the size of the model by one equation. The estimated coefficients are all consistent with expectations. It bears noting that the estimate of the marginal propensity to absorb (spend) out of increased income is 1.006. As Alexander has argued, it is not unreasonable for this coefficient to exceed unity because the marginal propensity to absorb includes induced investment.

In the interest rate equation the estimated coefficients of real income, lagged interest rate, and the rate of change in the inflation rate are all positive as expected. However, the estimated coefficient of the supply of high-powered money did not exhibit the expected negative sign. This may be due to the positive relationship between the rate of inflation and rate of interest. When the supply of high-powered money increases, it is expected to put downward pressure on the rate of interest. However, it also puts upward pressure on the rate of inflation which tends to raise the rate of interest. The magnitude of this coefficient estimate was so small that it would have very little effect on simulations of the model. Therefore, it was not dropped from the system.

The estimated Philips curve shows a strong positive relationship between changes in the stock of high-powered money and the wage rate. The unemployment rate shows a relatively weak negative effect on the rate of increase in the wage rate.

The net capital inflow equation also proved problematical. The estimated coefficients of the changes (first differences) in the rate of interest in the U.S. and in the foreign countries had the expected positive and negative signs, respectively. However, an attempt to include the respective rates of inflation as risk indicators failed to give the expected signs. These variables were therefore dropped from the equation.

In order to validate the model and evaluate its usefulness for simulating the effects of alternative shocks, the model was solved simultaneously over the sample period using a nonlinear solution package (Ragsdell) which employs Newton's method. For the interested reader, the detailed validation statistics are presented in Shei (pp. 93-99). Only the salient characteristics are summarized here in the interest of space. Overall, the fit of the model can be judged as quite acceptable. In the real sector the supply and absorption quantities and prices track well. The general price index, aggregate expenditures, and national income track very well. The weakest points are industrial exports and net capital inflows (and therefore the balance of payments) which can be judged as performing poorly. Crop exports and interest rate performance also leave much to be desired. Nevertheless, the overall performance of the model is judged as adequate for proceeding to illustrate its utility for policy analysis. For this purpose, it is employed to evaluate the various shocks to the system in accounting for the 1973 price increases.

DECOMPOSITION OF 1973 INFLATION

To decompose the observed 1973 price inflation, the 24-equation model is solved simultaneously under several alternative scenarios. The model is first solved for the years 1972 and 1973 using the observed values of all predetermined and exogenous variables. This gives the total changes in all endogenous variables predicted by the model, given all the shocks which actually occurred in 1973, which include: U.S. monetary expansion by $8.7 billion (i.e., in domestic component of the stock of high-powered money), dollar devaluation (cumulative) of 8.2 percent, a fall in world cereals production of 37 million metric tons, and a Soviet (and Chinese) grain purchase totaling almost $1 billion. To estimate the effects of these shocks, the model is solved again for 1973 with these four "variables" set at their 1972 levels to obtain the solution values of all endogenous variables which would have held if none of the four events had occurred. Then, the shocks are reintroduced one at a time to estimate the separate effects of each event as if none of the others had happened. The results for the variables of interest are summarized in Table 4.5.

Of particular interest are the effects of these four shocks on the general price level and their differential effects on prices in the four sectors. Of these four sources of the 1973 price inflation as measured by the general price level, domestic monetary expansion had the greatest effect, followed by the Soviet grain deal, dollar devaluation, and the fall in world grain production.[10] Note that agricultural price increases contributed more to the general price increase than did industrial and service prices.

Domestic monetary expansion accounts for 34 percent of the increase in crops prices, 60 percent of the livestock price rise, and almost half of the increases in industrial and services prices and in the general price level. These differences reflect, among other things, the differences in the marginal propensities to absorb the four product aggregates as expenditures increase. Nevertheless, one must bear in mind that the effects of these shifts are modified by the general equilibrium interactions among the four sectors as increases in prices in each sector in turn shift the supply schedules of other sectors, and so on until all the (short run) general equilibrium effects work themselves out.

Of the four shocks, the Soviet grain deal had the largest effect on domestic crops prices. It alone accounts for 57 percent of the observed increase in crops prices, and in turn for 20 percent of the increase in livestock prices. These shocks then work their way through the economy, raising the general price level, then the industrial and services wage rate and in turn prices in those sectors. This finally exerts some additional upward pressure on crop and livestock prices. The increase in export revenue causes monetary expansion, which also puts upward pressure on all prices in the economy. The net effect, as observed, accounts for over 13 percent of the increase in the general price level.

Dollar devaluation was the third largest source of increase in the general price level, accounting for only four percent of the observed increase. Similar to the other two shocks, its effect was

TABLE 4.5
The Price Effects of Four Shocks to the U.S. Economy, 1973

Shock	Crops Prices	Livestock Prices	Industrial Prices	Services Prices	General Price Level
	(Change in price index from 1972 to 1973)[a]				
Cumulative devaluation of 8.2 percentage points	2.39 (7.54)	1.02 (5.91)	0.38 (3.70)	0.31 (3.61)	0.41 (3.95)
Domestic monetary expansion of $8.7 billion	10.74 (33.91)	10.28 (59.69)	5.14 (49.76)	4.10 (48.52)	4.94 (48.03)
World cereal production decrease of 37 million M.T.	1.29 (4.06)	0.25 (1.42)	0.07 (0.69)	0.06 (0.67)	0.10 (0.95)
Soviet grain purchase of $963 million[b]	18.08 (57.11)	3.45 (20.00)	1.00 (9.67)	0.80 (9.48)	1.38 (13.39)

Total Explained	32.49 (102.63)	14.99 (87.02)	6.56 (63.82)	5.26 (62.29)	6.83 (66.33)
Unexplained	-0.83 (-2.63)	2.24 (12.98)	3.74 (36.18)	3.19 (37.71)	3.47 (33.67)
Change[c]	31.66 (100.00)	17.22 (100.00)	10.32 (100.00)	8.45 (100.00)	10.29 (100.00)

[a]In parentheses the values below each entry are the percents of the observed total change.

[b]The Soviet grain purchase in 1972/1973, together with Chinese purchases, accounted for about a third of the increase in the volume of U.S. grain exports in fiscal 1973 (Economic Report of the President 1974, p. 132). The observed increase in fiscal 1973 was $2.89 billion (price deflated). One-third of the observed increase is therefore, equal to $963 million.

[c]The "observed total" equals the difference between the base solution values for 1973 to 1972.

relatively greater on the prices of crops and livestock. It is interesting to note that while all prices increase as a result of dollar devaluation, the effect is greatest on crops prices, intermediate on livestock prices, and relatively smallest on industrial and services prices. These results suggest that the inflationary effects of dollar devaluation were transmitted most strongly through the crops sector, and secondarily through the livestock sector. Two effects are working here. First, the increase in the price of crops raises the livestock production cost structure, shifting its supply schedule upwards. At the same time, the expenditure elasticity of demand for livestock products is higher than for crops. In the general equilibrium analysis of the devaluation, the associated increase in real expenditures causes a larger proportional upward shift in the demand for livestock schedule than in the demand for crops schedule. Both shifts put upward pressure on livestock prices.

It is interesting to note that these general equilibrium results suggest smaller price effects of dollar devaluation than found by Schuh or Chambers and Just (1981) based on a more partial equilibrium framework in which intersectoral and monetary sector linkages were assumed away. The implications of this model can easily be compared with those of the standard partial equilibrium formulas (Thompson and Schuh) for calculating the price effects of market distortions. The coefficients in the model suggest an elasticity of excess supply from the United States of 1.4, the elasticity of excess demand used in the model was -1.0. Assuming the U.S. dollar had not been devalued, in 1973 it would have been 19.2% overvalued in relation to 1970 (SDR basis). If the devaluation had not occurred, the partial equilibrium formulae suggest that domestic crop prices would have been 8.0% lower and world market crop prices 11.1% higher. However, when the general equilibrium adjustments are also taken into account here, the cumulative devaluation from 1970 to 1973, only accounts for a 3.0% decrease in the domestic price and an 8.3% increase in export price (Shei, pp. 111-112). This suggests that the general equilibrium adjustments tend to blunt those implied by the partial equilibrium model. The 8.3% increase in export price should probably be considered an upper bound estimate because many general equilibrium adjustments in the rest of the world are omitted from this model.

The final simulation experiment concerned the effect of the observed 37-million metric ton decline in world cereals production in 1972. This accounted for less than one percent of the increase in the U.S. general price level, and only four percent of the increase in crops prices. This is partly explained by the fact that the U.S.

is not the only grain exporter, and that the effect of the shock is distributed over all countries which permit the world market price signals to be reflected into their domestic markets. It could also be argued that it is asking too much to expect one aggregate export demand schedule for the crops sector to adequately capture the effect on the U.S. of all the adjustments which occurred in world agricultural markets following the crop shortfalls (Thompson, pp. 4-14). Moreover, the effect of the Soviet grain purchase was already treated separately above. The important point to note, however, is that the effects of the shift in crop export demand associated with crop failure in the rest of the world were small relative to the effects of the U.S. monetary expansion, the Soviet grain deal and dollar devaluation.

To provide an indication of the overall performance of the model in accounting for the observed changes in price levels, the simulated effects of the four individual shocks are also summed up in Table 4.5 to obtain the overall effect. The proportion of the total change accounted for by the four individual shocks provides an indication of the explanatory power of the model. As indicated, 103 percent of the crop price increase and 87 percent of the livestock price increase are accounted for by the model. On the other hand, only 62-64 percent of the increase in service and industrial prices are explained. Some of the unaccounted for increase might be explained by omitted variables, including increases in petroleum prices which resulted from the OPEC oil embargo in the fall of 1973, the shifting character of U.S. domestic price controls, an upsurge in both domestic and foreign demand for industrial output, and unexpected capacity constraints in the U.S. industrial sector.

LIMITATIONS

It must be emphasized that the empirical work presented here is not represented by the authors as definitive. After reviewing recent developments in open economy macroeconomic and monetary theory, an attempt was made to synthesize the monetarist and structuralist approaches and empirically implement the resulting theoretical framework. The authors sought to do this in the simplest manner possible which would still reflect the important linkages among the real sectors and between the real and monetary sectors through which inflationary shocks are transmitted. While the model has firm theoretical foundations, certain simplifications made in the empirical implementation, together with weak empirical estimates of the parameters in several equations mean that the results presented should be treated with some caution by users who wish to draw policy inferences. Rather, the study should probably be regarded as suggestive of a potentially interesting and fruitful direction for U.S. agricultural sector analysis to take. It also suggests that there is much more work to be done in improving the empirical model to the point that one would wish to make policy recommendations based upon it.

Schuh and Chambers and Just have emphasized the exchange rate as the principal avenue through which monetary shocks impinge upon the agricultural sector. The theoretical framework above outlined demonstrates that under a floating exchange rate, this is one transmission path. However, during all but the last two years of the data series used here, the United States maintained fixed exchange rates, and even in those years, there existed a "dirty float" characterized by frequent government intervention to prevent the rate from moving as the markets were taking it. Furthermore, in light of our interest in accounting for the relative importance of the various forces that raised agricultural prices in 1973, we needed to be able to exogenously fix and parametrically vary the exchange rate, rather than treat it as endogenous to the system as did Chambers and Just (1982). Nevertheless, to make the model useful for understanding the monetary linkages to agriculture in the late 1970s and 1980s, it would be necessary to follow an approach such as Chambers and Just's and treat the exchange rate as an endogenous variable.[11 12]

In the empirical specification adopted here there are two principal links between the monetary and real sectors. The first is through aggregate expenditures, and the second is through the effect of changes in export revenue on the monetary base and in turn on the money supply. These have generally been ignored in other studies of inflation and agriculture. The results here are encouraging despite the fact that the empirical export demand and international capital flow equations were weaker than desirable. More work is needed to improve the quality of these empirical estimates in order to put much faith in the absolute magnitudes of the calculated effects. One problem with the specification used in the export demand equation was the assumption that domestic and traded goods are homogeneous and that therefore the law of one price holds. As a result, both export demand equations were estimated net of imports. The homogeneity assumption is probably unwarranted in the case of industrial goods and perhaps also for crops. Future work should treat foreign and domestic goods as less than perfect substitutes at this level of aggregation.

A related concern is that the two export demand relations treat all the rest of the world variables as exogenous. That is, it is assumed that no adjustments are induced in the rest of the world, e.g., in their national income, from shocks originating in the U.S. This is clearly unrealistic in light of the interrelatedness of the world economy and the large role of the United States therein. Nevertheless, to do anything else than assume these variables to be exogenous was clearly beyond the scope of this study.

Schuh (1980) has argued that under a fixed exchange rate regime macro policy affects agriculture mainly through the labor market (a higher unemployment rate slows down the structural transformation of the economy). Under flexible exchange rates he argues that the main effect is through the capital market instead. (He abstracts from the linkages through the goods markets and aggregate expenditures.) This model includes very simplistic equations to explain changes in the wage rate and interest rate in order to account for the changes in industrial and service sector prices and to generate international capital flows. No factor market detail is included in the real sector submodel. This is clearly a shortcoming which precludes addressing some very important questions concerning the interrelations between inflation and agriculture. Lamm has examined some of the linkages between agriculture and the macroeconomy with a model built around two-factor aggregate production functions for agriculture and manufacturing in which the labor and capital markets are explicitly included. Unemployment and investment are therefore endogenous to his system.[13] On the other hand, Lamm's is basically a closed-economy model which ignores exports and international capital flows. It also abstracts from the role of nontraded goods in affecting the response of an economy to monetary shocks. Therefore, Lamm's work suggests a more promising means of treating the transmission of inflationary shocks through the factor markets. Nevertheless, it abstracts from an important characteristic of the U.S. economy today--its openness to international trade and capital flows.

The model presented here is not strictly a general equilibrium model. The production point is not constrained to lie on the production possibility frontier, and the demand structure does not satisfy the required homogeneity, symmetry, and additivity. Clements has estimated a three-sector general equilibrium model of the U.S. economy including a monetary module in which all these conditions are forced to be satisfied by imposing restrictions on the parameters both within and across equations during estimation. He used full-information maximum likelihood estimation which takes account of all the restrictions simultaneously to give efficient and consistent estimates. The procedure has a great deal of appeal in terms of ensuring consistency throughout all the results. Nevertheless, if one accepts any of the structuralist arguments, the economy may not be operating on its production possibility frontier with all factors fully employed all the time. Industrial and service sector prices may not be determined in the same manner or react as quickly to shocks as agricultural sector prices. Monetary policy, in any case, does have real effects in Clements' model in the short run, although not in the long run (p. 470).

The three sectors in Clements' model produce exportables, importables, and nontradeables. Several individual product aggregates, including agricultural products as a separate class, would make the results much more interesting for our purposes. An attempt to examine inflation and agriculture within this context seem to be one potentially fruitful direction for future research.

Another important limitation of the approach presented here concerns its lack of attention to dynamics. Chambers and Just (1981, 1982) used quarterly data in their work. This seems like a more promising approach to address the dynamics of inflation and agriculture. An annual model will probably never be very useful in understanding the dynamics of the process. A more fundamental aspect of the dynamics of inflation concerns the manner in which price expectations are formed and in turn affect the transmission of inflation (Gordon, 1975). Lamm has paid considerable attention to this in his analysis, and greater attention to it is needed in future efforts to expand the present work.

In summary, numerous limitations of the empirical implementation of the theoretical framework presented in this paper have been identified. Each of the other models which contain explicit linkages between agriculture and the macroeconomy helps clarify some of the relationship between inflation and agriculture. However, no one model in the literature to date puts all the pieces together. The challenge which remains is to capture the essence of the most important linkages in one model which tracks and forecasts well, but which is still small enough to be manageable.

CONCLUSIONS AND IMPLICATIONS

A basic premise of this paper is that the money supply has been an important omitted variable in past interpretations of the relation between agriculture and inflation in the United States. The evidence presented here suggests that the money supply is a variable which has had a significant effect on the general price level, and that changes in the money supply alter agricultural prices relatively more than prices in other sectors of the economy within a given time period. The evidence presented suggests that the factor which accounted for the largest single share of the rapid 1973 agricultural and general price inflation in the U.S. was the 10% expansion in the domestic component of the monetary base in that year. This is not to argue that monetary policy was the only cause of the observed price increases, but rather that it has been an important omitted variable in most explanations offered and that it is difficult to fully account for agricultural price behavior in the 1970s without taking it into account.

The theoretical framework developed in this paper provides a synthesis in one unified model of the various structuralist and monetarist explanations of the behavior of agricultural prices and the general price level. Most explanations of price behavior in the early 1970s have been based upon structuralist arguments, which have ignored recent advances in the monetarist approach. It is shown that the recent fixprice-flexprice models make structuralist assumptions about the nature of the real sector, while accepting that some changes in the money supply may be autonomous. Arguing that agricultural prices are more flexible than industrial prices, this implies that agricultural prices adjust relatively more in any given time period--on both the up side and the down side. This suggests that while monetary expansion in the early 1970s pushed up agricultural prices relatively more than prices in other sectors within a given time period, the early 1980s' policy of monetary restriction can be expected to have lowered agricultural prices relative to those in other sectors. This is consistent with what we have observed. This suggests that at times agricultural price policy and monetary policy may work at cross purposes to one another. A time of budget cutting and monetary stringency may so depress relative agricultural prices that government expenditures to support farm prices must increase.

Those few analysts who recognized that monetary policy has an important role in accounting for observed agricultural price behavior have tended to view the exchange rate as the only path through which its effects are transmitted to the agricultural sector. Drawing upon recent developments in the monetary approach to the balance of payments, it was demonstrated that aggregate expenditures provide a more direct linkage through which monetary shocks affect agriculture. If one considers the exchange rate as the only channel of transmission, one will understate the effect of monetary policy on agricultural prices. It is also demonstrated that partial equilibrium analysis of the effects of dollar devaluation on agriculture which omit intersectoral linkages and linkages between the real and monetary sectors, overstate the effect of the devaluation on domestic agricultural prices. The general equilibrium adjustments tend to blunt the effects predicted by partial analysis, although the exchange rate still has a significant effect.

NOTES

An earlier abbreviated version of this paper was presented at the 1979 annual meetings of the American Agricultural Economics Association (Shei and Thompson, 1979).

1. Here we view "inflation" as a once and for all rise in the price level or a fall in the purchasing power of money. As Lipsey points out, "Sometime in the 1970s...the definition switched to distinguish a change in the price level defined as a once-for-all change and a 'sustained' change in the price level, defined as inflation" (p. 285). The notion of continually rising prices or of persistent upward movement in the general price level has tended to replace any rise in the general price level. We concur with Lipsey that "The currently accepted definition is so vague that one must wonder how it came to replace the very much more precise definition that inflation is a rise in the price level" (p. 285). Our usage in this paper will be according to the old definition.

2. The recent analysis has taken on more of a micro focus, emphasizing what inflation "does" to agriculture. This recent literature, which has been reviewed by Prentice and Shertz, treats the effects of inflation on resource allocation, investment, prices paid relative to prices received, farm size, and concentration of wealth. The passthrough from farm prices to retail food prices has also been studied (Lamm). Most of these studies treat agriculture as a passive actor within the U.S. economic system, being affected by events in the rest of the system, but too small to affect them. In this study we abstract from such micro issues, not because they are unimportant, but because we focus on a different set of issues which have tended to be glossed over in previous discussions of inflation and agriculture. Here we emphasize the simultaneous determination of agricultural prices and prices in other sectors, the weighted

average of which is the general price level. We emphasize the linkages through which the various sectors interact and through which inflationary shocks get transmitted.

3. Most economists now acknowledge that the general price index cannot continue increasing for long without an increase in the money supply. The basic question is one of the direction of causality. The structuralists argue the monetary expansion merely accommodates price rises which originate in the real sector (Olivera). Monetarists argue that monetary expansion is an autonomous causal force.

4. Bordo argues that the response of relative prices to shifts in demand caused by changes in the money supply is associated with the length of wage and price contracts. He argues that the length of a fixed price contract negotiated by an industry, for given transactions costs and risk aversion, is inversely related to the variance of the industry's relative prices. The shorter the contract within an industry, the more flexible will be its price, and the more responsive its price will be to a given monetary change.

5. In their analysis of the "1973 food price inflation" Eckstein and Heien assert that they examine the effects of each of these and conclude that domestic monetary policy was the most important explanatory variable. However, no explicit link between money supply and agricultural sector variables is included in their econometric model, which is basically a U.S. grains-livestock sector model.

6. For example, following Dornbusch, the domestic price of nontradeable goods is determined solely by domestic supply and demand conditions. The effect of any given shock on nontradable goods prices is determined to a considerable extent by how substitutable they are in supply and demand for exportables and importables.

7. In the strictest sense of the word this is not a general equilibrium model. Strictly speaking, to be called a general equilibrium model, a model should exhibit two characteristics which this specification does not. First, the economy's production point should be constrained to lie on the production possibility frontier, the position and shape of which are dictated by the economy's factor endowments and technology (production function). Second, the structure of consumer demand should satisfy the three demand-theoretic restrictions of homogeneity, symmetry of the substitution effects, and additivity. Clements has constructed a three-sector econometric model of the U.S. economy including a monetary sector which satisfies these conditions. However, his three sectors are

importables, exportables, and nontraded goods. The sectoral identity, in which we are interested gets lost in his model. Moreover, no structuralist characteristics of the economy can be included in such a specification. Lamm builds a two-sector model around the production functions for the agricultural and manufacturing sectors in which the means by which price expectations are formulated is treated in considerable detail.

8. Over the period 1970-74, U.S. agricultural exports and imports accounted for 14.7 and 8.2 percent respectively, of the total value of world agricultural exports and imports. U.S. industrial exports and imports represented 10.2 and 10.7 percent, respectively, of total world industrial goods exports and imports in the same period.

9. The U.S. dollar is the "key currency," and dollar holdings represent a large fraction of the international reserves held by many countries. Therefore, the distinction between international reserves and domestic money gets somewhat blurred in the U.S. case.

10. Eckstein and Heien, using a set of econometric models of only the U.S. livestock and grains sectors reached a similar ranking of the major causes of the 1973 food price inflation: domestic monetary policy, government acreage restrictions, the Soviet grain deal, world economic growth, dollar devaluation, and the price freeze. It should be emphasized, however, that no explicit monetary sector was included in their model.

11. There are those who would also argue that the money supply itself should also be treated as an endogenous variable. Politico-economic models which have been formulated to explain the behavior of the Fed include variables contained in this model as arguments of the Fed's "reaction function." For example, Gordon (1975) argues that the equilibrium change in the money supply is determined by the supply and demand for inflation. Pressure groups "demand" an increased money supply to accommodate price increases in certain sectors if their jobs would be lost or income or wealth fall from the necessary reduction that would occur in other prices if the monetary authority held the line on the money supply. The "supply" of inflation reflects the extent to which the government bows to such pressures. This depends to no small extent on the expected future election losses from resistance.

12. Soe Lin took a similar approach to Chambers and Just. However, instead of building his own simplified monetary sector model, he employed the very detailed RDX2 model of the Bank of Canada to generate the exchange rate and general price level, which were in turn fed into his econometric model of the Canadian

agricultural sector. It is frequently argued that this is the approach taken by the private econometric forecasting firms which maintain both agricultural sector and macroeconomic models of the United States. However, due to their proprietary nature it is not possible to document their experience with this approach in practice.

13. Recall that investment is subsumed under absorption in the model presented above. For an alternative to Lamm's approach, Hughes specified and estimated a general equilibrium model of U.S. agriculture in which considerable financial market detail is included.

REFERENCES

Alexander, S. S. "The Effects of a Devaluation on a Trade Balance." IMF Staff Papers. 2(1952):263-278.

Barnett, R. C., D. A. Bessler, and R. L. Thompson. "Monetary Changes and Relative Commodity Prices: A Time-Series Analysis." Processed. Department of Agricultural Economics, Purdue University, September 1981.

______________. "The Money Supply, and Nominal Agricultural Prices." American Journal of Agricultural Economics. 65(1983): 303-307.

Baumol, W. J. "Macroeconomics of Unbalanced Growth: The Anatomy of Urban Crisis," American Economic Review 57(1967):415-426.

Bordo, M. D. "The Effects of a Monetary Change on Relative Commodity Prices and the Role of Long Term Contracts," Journal of Political Economy. 88(1980):1088-1109.

Bryant, R. C. "Empirical Research on Financial Capital Flows," International Trade and Finance--Frontiers for Research. P. B. Kenen (ed.). Cambridge, U.K.: Cambridge University Press, 1975. pp. 321-362.

Chambers, R. G., and R. E. Just. "Effects of Exchange Rate Changes on U.S. Agriculture: A Dynamic Analysis," American Journal of Agricultural Economics. 63(1981):32-46.

______________. "An Investigation of the Effect of Monetary Factors on Agriculture," Journal of Monetary Economics. (1982):235-247.

Clark, P. B. "The Effects of Recent Exchange Rate Changes on the U.S. Trade Balance," The Effects of Exchange Rate Adjustments. P. B. Clark, et. al. (ed.), Department of the Treasury, Washington, D.C., 1974. pp. 201-236.

Clements, Kenneth W. "A General Equilibrium Econometric Model of the Open Economy," International Economic Review. 21(1980): 469-488.

Dornbusch, R. "Devaluation, Money, Nontraded Goods," American Economic Review. 63(1973):871-880.

Eckstein, A., and D. Heien. "The 1973 Food Price Inflation," American Journal of Agricultural Economics. 60(1978):186-196.

Frenkel, J. A., and H. G. Johnson, (eds.). The Monetary Approach to the Balance of Payments. Toronto: Univ. of Toronto Press, 1976.

Gordon, R. J. "Recent Development in the Theory of Inflation and Unemployment," Journal of Monetary Economics. 2(1976):185-219.

__________. "The Brookings Model in Action: A Review Article," Journal of Political Economy. 78(1970):489-525.

__________. "The Demand for and Supply of Inflation," Journal of Law and Economics. 18(1975):807-836.

Griliches, Z. "The Demand for Inputs in Agriculture and a Derived Supply Elasticity," Journal of Farm Economics. 38(1959):309-322.

Hathaway, D. E. "Food Prices and Inflation," Brookings Papers Economic Activity. (1974):63-116.

Hooper, P. "An Analysis of Aggregation Error in U.S. Merchandise Trade Equations." Ph.D. thesis, Univ. of Michigan, 1974.

Hooper, P., and J. F. Wilson. "Two Multi-Level Models of U.S. Merchandise Trade, 1958.I-1971.IV, and Post Sample Analysis, 1972.I-1973.II: An Evaluation of a Workable Forecasting System." Discussion Paper No. 47, Division of International Finance, Board of Governors of the Federal Reserve System, Washington, D.C., 1974.

Houthakker, H. S. "The 1972-75 Commodity Boom: Comments," Brookings Paper Economic Activity. 3(1975):716-717.

Houthakker, H.S., and S. P. Magee. "Income and Price Elasticities in World Trade," Review of Economics and Statistics. 51(1969):111-125.

Hughes, D. W. "A General Equilibrium Model of Agriculture as Part of the U.S. National Economy." Ph.D. thesis, Texas A & M University, 1980.

Johnson, H. G. "The Keynesian Revolution and the Monetarist Counterrevolution." American Economic Review. (Papers & Proceeding) 61(1971)1-14.

Johnson, P. "The Elasticity of Foreign Demand for U.S. Agricultural Products," American Journal of Agricultural Economics. 59(1977):735-736.

Khan, M. "Inflation and International Reserves: A Time Series Analysis," IMF Staff Papers. 26(1979):699-724.

Lamm, R. M., Jr. "The Role of Agriculture in the Macroeconomy: A Sectoral Analysis," Applied Economics. 12(1980):19-35.

Lawrence, R. Z. "Primary Commodities and Asset Markets in a Dualistic Economy." Paper presented at USDA/Universities Consortium for Agricultural Trade Research Conference on Macroeconomic Linkages to Agricultural Trade, Tucson, AZ, Dec. 15-17, 1980.

Lipsey, R. G. "World Inflation," Economic Record. 151(1979):283-96.

Magee, S. P. "A Theoretical and Empirical Examination of Supply and Demand Relationships in U.S. International Trade." Study prepared for the Council of Economic Advisers, Washington, D.C., October 1970.

Mundell, R. A. International Economics. (New York, Macmillan, 1968).

Olivera, J. H. G. "On Passive Money," Journal of Political Economy. 78(1970):805-814.

Prais, S. J. "Some Mathematical Notes on the Quantity Theory of Money in an Open Economy," IMF Staff Papers. 8(1961):212-226.

Prentice, P. T. and L. P. Schertz. "Inflation: A Food and Agricultural Perspective," AER No. 463, Economics and Statistics Service, U.S. Department of Agriculture, Washington, D.C. 1981.

Ragsdell, K. M. MELIB: A Tool for Mechanical Engineers. Mimeo. Department of Mechanical Engineering, Purdue University, Jan. 1975.

Sayad, J. "Inflacao e Agricultura," Pesquisa e Planejamento Economico. 9(1979):1-32.

Schnittker, J. H. "The 1972-73 Food Price Spiral," Brookings Paper Economic Activity. (1973):498-507.

Schuh, G. E. "The Exchange Rate and U.S. Agriculture," American Journal of Agricultural Economics. 56(1974):1-13.

__________. "Income Stability: Implications of Monetary, Fiscal, Trade, and Economic Control Policies," Farm and Food Policy Symposium. J. S. Plaxico, (ed.). Great Plains Agricultural Council Publication No. 84, September 1977, pp. 69-96.

__________. "Floating Exchange Rates, International Interdependence, and Agricultural Policy," Rural Change--The Challenge for Agricultural Economists. G. Johnson and A. Maunder, (eds.). Westmead, England: Gower, for the International Association of Agricultural Economists, 1981, pp. 416-423.

Shei, S-Y. "The Exchange Rate and United States Agricultural Product Markets: A General Equilibrium Approach." Ph.D. thesis, Purdue University, 1978.

Shei, S-Y, and R. L. Thompson. "Inflation and U.S. Agriculture: A General Equilibrium Analysis of the Events of 1973." Contributed Paper, Annual Meetings of the American Agricultural Economics Association, Washington State University, 1979.

Sims, C. "Money, Income and Causality," American Economic Review. 62(1972):540-552.

Soe-Lin. "A Macro Policy Simulation Model of the Canadian Agricultural Sector." Ph.D. thesis, Carleton University, 1980.

Thompson, R. L. "A Survey of Recent U.S. Developments in International Agricultural Trade Models." Bibliographies and Literature of Agriculture, No. 21, Economic Research Service, U.S. Department of Agriculture, Washington, D.C. 1981.

Thompson, R. L. and G. E. Schuh. "Trade Policy and Exports: The Case of Corn in Brazil," Pesquisa e Planejamento Economico. 8(1978):663-694.

Tweeten. L. G. "The Demand for United States Farm Output," Food Research Institute Studies. 7(1967):343-369.

Tweeten, L. G. and Quance, C. L. "Positivistic Measures of Aggregate Supply Elasticities: Some New Approaches," American Journal of Agricultural Economics. 51(1969):342-352.

Van Stolk, A. P. "World Grain Economy--Price Effect on Demand," Wheat Production Potentials and Trade Patterns: A New Era?. Proceedings of 1976 International Seminar on Wheat, Washington, D.C.: Great Plains Wheat, Inc., 1976, pp. 19-25.

Wacher, M. "The Changing Cyclical Responsiveness of Wage Inflation," Brookings Paper Economic Activity. (1976):115-167.

5
Overshooting of Agricultural Prices

Kostas G. Stamoulis and
Gordon C. Rausser

INTRODUCTION

A review of the agricultural economics literature over the past 15 years reveals a growing interest in the effects of macroeconomic aggregates, especially monetary instruments, on the U.S. farm sector. The possibility that the prosperity of U.S. agriculture in the early 1970s and its demise during the early 1980s could be linked to major changes in the macroeconomic environment during those periods induced several researchers to take a closer look at the interaction between agriculture and the rest of the U.S. economy.

The high correlation between degree of "easiness" in monetary policy and the behavior of relative prices of farm products during the early 1970s, and again in the late 1970s, appears to support this view. The expansion of the U.S. money supply to accommodate oil price increases has been associated with a dramatic increase in real commodity prices. In contrast, the squeeze in the credit markets resulting from a tight money supply and the high budget deficit during the early 1980s is associated with a depression in real agricultural prices and incomes.

Although casual observation suggests a relationship between money supply and relative farm prices, two other sources of impact on the agricultural sector should be recognized before the effects of

---

Kostas G. Stamoulis is Assistant Professor, Department of Agricultural Economics, University of Illinois at Urbana-Champaign, and Gordon C. Rausser is Robert Gordon Sproul Chair Professor of Agricultural and Resource Economics, University of California at Berkeley.

monetary policy can be accurately assessed: (1) exogenous shocks to demands and supplies of agricultural commodities not related to macroeconomic policy and (2) public policy directed at the farm sector. Exogenous shocks during the early 1970s include the move to flexible exchange rates and the ensuing dollar devaluation, the decrease in barriers between agriculture and other sectors of the economy, and the growth in demand for U.S. exports by less developed and communist countries; all could be associated with increases in real farm prices. At the same time, agricultural policy led to the elimination of the large government grain stocks (accumulated during the 1960s) which also contributed to the increase in real farm prices.

Similarly, the experience of the early 1980s could be associated with several other factors related to (1) and (2) above, besides the tight monetary policy environment. One important exogenous factor is the effort of foreign governments to support farm income which resulted in subsidies that encouraged production increases and caused export markets for U.S. products to shrink. And U.S. agricultural policy in the 1980s brought large income transfers to the farm sector through deficiency payments combined with relatively ineffective supply control schemes which are at least partially responsible for several record crops.

While the stylized facts establish a correlation between money and relative farm prices, no agreement exists on the significance of this relationship. Empirical results range from a significant relationship between the size of the money supply and real commodity prices (Chambers and Just, 1982) to no relationship at all (Batten and Belongia, 1984). Part of the explanation for these opposite outcomes are alternative theoretical macroeconomic paradigms which imply quite different price behaviors. Within a strict monetarist framework, for instance, monetary changes should have no effect on relative prices either in the short or long run.

It is important to stress relative prices for two reasons: (1) A policy that leaves relative prices of agricultural products unchanged is of no interest from a policy perspective. (2) Admitting monetary policy effects on relative prices requires identification of the special characteristics that separate or distinguish the farm sector from other sectors of the economy. Therefore, a model is needed that distinguishes agriculture from other sectors so that the effects of monetary policies on the farm sector can be isolated.

In what follows, a theoretical model is constructed that allows the separation between fix-price and flex-price markets. Agriculture is assumed to be a flex-price sector while manufactures and services are assumed to be sticky-price markets. But this classification is not tight. There are nonagricultural markets characterized by price flexibility, and there are cases in which farm prices exhibit downward stickiness (e.g., when supported prices for grains constitute price floors).

Despite these exceptions, the above distinction underlines the basic characteristics of agricultural and nonagricultural markets. In agriculture, day-to-day trading, widely disseminated information, and use of several agricultural commodities as financial assets make prices sensitive to changes in demands, supplies, and expectations. In manufactures and services, long-term contracts and costly adjustments to changes in market conditions, limit price responsiveness.

In a world in which all prices are flexible, a monetary shock is instantly translated into proportional changes of prices in all markets--leaving relative prices unchanged. Money neutrality, in that case, holds for both the short run and the long run. But if some prices are sticky, a change in nominal money is also a change in real money. For an increase in the money supply, interest-rate reductions cause portfolio shifts between storable commodities, financial assets, and currency. Under certain conditions, the prices of commodities overshoot their long-run equilibrium. Using a variant of Dornbusch's (1976) model of exchange-rate fluctuations, we show that, following a monetary shock, prices of agricultural commodities may overshoot their long-run equilibrium if prices in the rest of the economy are sticky.

In Section 2, the basic price-adjustment model is presented along with the conditions necessary for overshooting. By relaxing assumptions in the basic model, it is shown that, even in the presence of inflexible prices for some sectors, overshooting of flexible prices may not occur if certain conditions hold. Thus, whether overshooting occurs or not becomes an empirical question. In Section 3, empirical evidence on differential price adjustment to changes in money growth is presented. Brief reviews of past findings are included along with some new empirical evidence. In Section 4, concluding remarks are made and some policy implications are given.

THE BASIC MODEL

The theoretical model is a variant of Dornbusch's (1976) overshooting model constructed to explain movements in flexible exchange rates. In Dornbusch's model, the prices of all goods are assumed to be sticky, adjusting less rapidly than asset prices. While our analysis also focuses on exchange rate fluctuations, Dornbusch's model is altered to include a flexible commodity market. In a similar analysis, Frankel and Hardouvelis (1985) expressed the model in terms of prices of commodities and manufactures.

As in Dornbusch (1976), we assume that uncovered interest parity holds; i.e.,

$$r = r^* + x \tag{1}$$

for

r = the domestic short-term nominal interest rate
r^* = the foreign short-term nominal interest rate
x = the expected rate of depreciation or appreciation of the domestic currency where the exchange rate (E) is defined as the domestic currency price of the foreign country's currency (for the case of the U.S., dollars per unit of foreign currency).

In this simplified version of the model, we consider the "domestic" country to be small which implies that the nominal interest rate adjusted for expected depreciation equals the (given) foreign rate. Implicit in the equation are the assumptions of perfect substitutability between domestic and foreign interest-bearing instruments (one-bond world), absence of risk premia, and perfect capital mobility.

The expected rate of depreciation or appreciation is defined as being proportional to the gap between the exchange rate and its long-run equilibrium value (bars denote long-run values)

$$x = \theta(\bar{e} - e), \qquad \theta > 0 \tag{2}$$

where e is the logarithm of the exchange rate. The above regressive expectations scheme simply says that if the spot rate exceeds its long-run value, which is assumed known, then investors expect the rate to gradually appreciate at a speed of adjustment equal to θ. In the long-run, and in the absence of disturbances, $e = \bar{e}$ and $x = 0$. Equations (1) and (2) together imply that, for long-run equilibrium, $r = r^*$.

For the money market, a standard money demand equation is assumed (Except for r, lower case letters denote logs.):

$$m - q = \phi y - \lambda r \tag{3}$$

m = money supply
q = real composite price index described below
y = real output
λ = the interest rate semi-elasticity of demand for real balances.

We construct q on the assumption that the economy consists of two sectors (goods)--a flexible-price good and a fix-price good. For the flex-price good (presumably an agricultural or other primary commodity), Purchasing Power Parity (PPP) holds both in the long and the short run. In other words, if by P_A we denote the price of the flex-price commodity and let P_A^* denote its foreign counterpart, the PPP simply says that (in logs):

$$e = p_A - p_A^*. \tag{4}$$

The small country assumption permits us to set an arbitrary value for P_A^*. By setting $P_A^* = 1$, then $p_A^* = \log(P_A^*) = 0$ and equation (4) simply becomes:

$$p_A = e. \tag{5}$$

To construct q, we assume that the underlying utility functions are Cobb-Douglas so that construction of the price index requires that prices for the two commodities be weighted by their expenditure shares:

$$Q = (P_N^{\alpha})\, P_A^{(1-\alpha)} \tag{6}$$

and in log form:

$$\begin{aligned} q &= \alpha p_N + (1-\alpha)\, p_A \\ &= \alpha p_N + (1-\alpha)\, e \quad \text{for } 0 \leq \alpha \leq 1 \end{aligned} \tag{7}$$

where P_N is the price of the fix-price good and α is the nonagricultural expenditure share.

Taking into account the whole price index, real-money demand becomes:

$$m - \alpha p_N - (1-\alpha)e = \phi y - \lambda r. \tag{8}$$

Combining equations (1), (2), and (8), gives:

$$m - \alpha p_N - (1 - \alpha)e = - \lambda\theta(\bar{e} - e) - \lambda r^* + \phi y. \qquad (9)$$

At this stage, we assume a stationary money supply which implies that $m = \bar{m}$ and that interest rates are equalized in the long-run ($r = r^*$).

Thus, the long-run version of equation (9) becomes:

$$\bar{m} - \alpha\bar{p}_N - (1 - \alpha)\bar{e} = - \lambda r^* + \phi y. \qquad (10)$$

From equations (9) and (10) substituting $- \lambda r^* + \phi y$ from equation (10) and assuming that output is fixed (i.e., $y = \bar{y}$), we get:

$$\alpha(\bar{p}_N - \bar{p}_N) + (1 - \alpha)\ (\bar{e} - e) = - \lambda\theta(\bar{e} - e)$$

or (11)

$$e = \bar{e} - \alpha[1 - \alpha) + \lambda\theta]^{-1}(p_N - \bar{p}_N).$$

OVERSHOOTING

Equation (11) states that the spot rate deviates from its long-run equilibrium value by an amount proportional to the deviation of sticky sector from their long-run equilibrium values. The factor of proportionality, $\alpha/[(1-\alpha) + \lambda\theta]$, depends positively on the relative weight of sticky prices in the price index while it is a decreasing function of the relative weight of flexible prices. From (11), by differentiating with respect to m and noting that $dp_N/dm = 0$ (from the short-run stickiness assumption) and also that $d\bar{p}_N/dm = d\bar{e}/dm = 1$ (long-run neutrality), we find:

$$\frac{de}{dm} = 1 + \frac{\alpha}{(1 - \alpha) + \lambda\theta} > 1, \qquad (12)$$

i.e., the exchange rate overshoots its long-run equilibrium following a monetary change. For extreme values of α, we obtain the following:

For $\alpha = 1$, $\frac{de}{dm} = 1 + \frac{1}{\lambda\theta}$

which is the result reached by Dornbusch, and

for $\alpha = 0$, $\frac{de}{dm} = \frac{d\bar{e}}{dm} = 1$

which is to be expected since, with all prices flexible, a monetary shock causes all prices to return instantly to their long-run equilibrium.

To derive the "overshooting coefficient," $\alpha/[1-\alpha) + \lambda\theta]$, in terms of the parameters of the model, we need to solve for the coefficient of adjustment (θ) of sticky prices. Again, assuming output to be constant in the short-run, we specify the rate of change of sticky prices to be proportional to the gap between real output and aggregate demand, i.e., we assume disequilibrium in the fix-price markets in which prices adjust to changes in the "inflationary gap" (see also Dornbusch). Aggregate demand for output is thus defined as:

$$\ln(D) = u + \delta(e - p_N) + \gamma y - \sigma r \tag{13}$$

where $e - p_N$ is the relative price of domestic output. Excess demand is defined as the difference between actual and potential income or by $(\ln(D) - y)$. Thus, positive excess demand exerts an upward pressure on prices of both inputs and outputs, thereby increasing the price level, while the opposite holds for slack demand.

We could then express the rate of change in sticky prices as:

$$\begin{aligned}\dot{p}_N &= \pi[u + \delta(e-p_N) + \gamma y - \sigma r - y] \\ &= \pi[u + \delta(e-p_N) + (\gamma - 1)y - \sigma r].\end{aligned} \tag{14}$$

Here π is the rate at which sticky prices adjust to the gap between real output and demand, and dots over denote time rates of change. Substituting r with $r^* + \theta(\bar{e} - e)$, and recalling that $\bar{e} - e = \alpha[(1-\alpha) + \lambda\theta]^{-1}(p_N - \bar{p}_N)$ from (11), we have:

$$\dot{p}_N = -\pi\left[\frac{\delta(1 - \lambda\theta) + \alpha\sigma\theta}{(1 - \alpha) + \lambda\theta}\right](p_N - \bar{p}_N). \tag{15}$$

Differentiating (11) with respect to time and substituting $\dot{p}_N$ from (15), we obtain:

$$\dot{p}_A = \dot{e} = \frac{\alpha}{[(1-\alpha)+\lambda\theta]}\left[\frac{-\pi[\delta(1+\delta\theta) + \alpha\sigma\theta]}{[(1-\alpha) + \lambda\theta]}\right](p_N - \bar{p}_N) \tag{16}$$

By rearranging (11), $(p_N - \bar{p}_N) = \frac{1}{\alpha}[(1-\alpha) + \lambda\theta](\bar{e} - e)$, and thus (16) becomes:

$$\dot{p}_A = \dot{e} = \pi\left[\frac{\delta(1 + \lambda\theta) + \alpha\sigma\theta}{(1 - \alpha) + \lambda\theta}\right](\bar{e} - e) \tag{17}$$

which shows the actual rate of depreciation of the currency and actual rate of growth of p_A. The expected rate of depreciation x is given by (2) as

$$x = \theta(\bar{e} - e).$$

Perfect foresight expectations requires that both actual and expected rates of depreciation be equal. Thus, the perfect foresight speed of adjustment (θ) of the system to its long-run equilibrium can be derived from the solution of the quadratic equation:

$$\theta = \pi\left|\frac{\delta(1 - \lambda\theta) + \alpha\sigma\theta}{(1 - \alpha) + \lambda\theta}\right| \tag{18}$$

From (18), solving for θ,

$$\theta = \pi(\lambda\delta + \alpha\sigma) - (1-\alpha)) \pm \frac{\sqrt{[(1-\alpha) - \pi(\lambda\delta + \alpha\sigma)]^2 + 4\lambda\pi\delta}}{2\lambda} \tag{19}$$

Substitution of (19) into (12) gives the overshooting coefficient in terms of the parameters of the model.

The above model shows that, in a world in which some prices are sticky, the burden of adjustment to a monetary shock is borne by the flexible price sectors. It is also worthwhile to notice that short-run nonneutrality of money holds even though agents have perfect foresight about future price paths.

OVER- VS. UNDERSHOOTING

It is possible to demonstrate that the fix-price, flex-price separation is a necessary but not sufficient condition for flexible prices to overshoot their long-run equilibrium. Specifically, if we relax the assumption of a fixed real output, it can be shown that the prices in flexible markets may undershoot their long-run equilibrium values following a monetary shock [see also Dornbusch (1976), Appendix]. In this version of the model, all the assumptions of section 2.1 hold except that real income is assumed to be sensitive to changes in interest rates over the short-run. Thus, the goods markets clear in the short-run, although short-run equilibrium output and prices are different than long-run ones. It is exactly this difference that causes p_N to change. Under the new assumption, (13) becomes:

$$\ln (D) = y + u + \gamma y - \sigma r + \delta(e - p_N) \tag{20}$$

or

$$y = (\frac{1}{1 - \gamma})[u - \sigma r + \delta(e - p_N)]. \tag{21}$$

Prices adjust at a rate π proportional to the difference between short-run and long-run output:

$$\dot{p} = \pi(y - \bar{y}). \tag{22}$$

Equations (21) and (22) summarize the adjustment to long-run equilibrium of output and prices as a series of short-run equilibria. In the long-run, $q = \bar{q}$, $\dot{p}_A = \dot{p}_N = 0$, $y = \bar{y}$. Following a solution process similar to the one in the previous case, the key relationships of the model become:

$$(e - \bar{e}) = - \left[\frac{\alpha - \phi\delta\mu}{[(1-\alpha) + \phi\mu(\delta+\sigma\theta) + \lambda\theta]}\right] (p_N - \bar{p}_N) \tag{23}$$

where

$$\mu = \frac{1}{1 - \gamma}.$$

and α, δ, ϕ, θ, and σ are as defined in the previous section; Differentiating 23,

$$\frac{de}{dm} = 1 + \frac{\alpha - \phi\delta\mu}{(1-\alpha) + \phi\mu(\delta + \sigma\theta) + \lambda\theta}. \tag{24}$$

Since the denominator of (24) is always positive, overshooting occurs when $\alpha - \phi\delta\mu > 0$ or $\alpha > \phi\delta\mu$. Equation (24) shows that (for positive θ) the speed of adjustment of prices does not determine whether or not overshooting occurs but the <u>extent</u> to which flexible prices overshoot their long-run equilibrium.

Comparison of (12) and (24) reveals that the additional terms in (24) summarize the output response to short-run changes in relative prices and the interest rate (δ , σ) as well as the changes in the demand for money and goods caused by the change in output (ϕ, μ).

The perfect foresight solution for θ (the coefficient of adjustment of prices) is determined by solving the quadratic equation:

$$\theta = \pi \left[\frac{\frac{1}{1-\gamma}[\delta(1 + \lambda\theta) + \alpha\sigma\theta]}{(1 - \alpha) + \phi\mu(\delta + \sigma\theta) + \lambda\theta} \right]. \tag{25}$$

An intuitive explanation of the results in sections 2.1 and 2.2. is as follows: A shock in nominal money that leaves part of the price level unchanged is a shock in real money. The interest rate falls to clear the asset markets that are assumed to be always in equilibrium. The exchange rate depreciates because of the incipient capital outflow and p_A, the price of the commodity, rises instantly due to the PPP condition. Depreciation continues until the expectation of future appreciation justifies the domestic-foreign nominal interest rate disparity. The fall in the interest rate causes demand for domestic output to rise while the fall in the value of the currency causes export demand to rise. As a result, aggregate demand rises along with real output causing sticky prices to rise gradually. As prices rise, real money balances fall, the interest rate rises, and the process is reversed for e and p_A until the system returns to long-run equilibrium

The basic difference between the two cases of overshooting and (possible) undershooting [summarized in (12) and (24)], is the response of output to interest rate changes and the subsequent effects on the demand for real balances. As output rises following a drop in the interest rate, part of the increase in the supply of money is absorbed, thus reducing the excess supply of money.

A smaller excess supply of money means that a smaller decrease in the interest rate is needed to clear the asset markets. Consequently, the initial domestic-foreign interest rate gap, the currency depreciation, and the rise in p_A will all be smaller than in the case of fixed short-run output.

In summary, the overshooting parameter depends on both the income elasticity of money demand and the interest elasticity of aggregate output. From (24), as ϕ and σ rise, the overshooting parameter falls. Given the share of sticky prices in the price index (α), a positive monetary shock can cause undershooting of flex-prices if ϕ and α are sufficiently large.[1]

Following the exchange-rate literature, similar models could be constructed in which flexible prices may not overshoot because of wealth effects (Driskill, 1980; Engel and Flood, 1985) or because of sluggish capital mobility, imperfect bond substitutability, etc.[2] It is also possible to imagine situations in which overshooting of the exchange rate does not necessarily imply commodity price overshooting. Grain prices in the United States constitute a good example. Given the structure of the farm sector policy in the United States until 1985, in some instances support prices constituted a lower limit for grain prices. Thus, a drop in the supply of money may not cause grain prices to overshoot downward to the extent that they are close to or at support rates before the shock occurred. In general, any reason that would cause the purchasing power parity assumption to be violated will imply that exchange-rate overshooting will not imply overshooting of commodity prices (Obstfeld, 1986).

PRICE STICKINESS, MONETARY POLICY, AND RELATIVE PRICE CHANGES

Although the possibility of overshooting of commodity prices could have important implications for commodity-price variability, the single most important implication of the overshooting model is that it provides the theoretical basis for examining the existence of short-run real effects of money and monetary policy on the agricultural sector. While neutrality of money in the long-run is widely accepted, less agreement exists among economists as to the short-run relationship between money and relative prices. In his survey of price adjustment studies, Gordon (1982) considers the short-run inertia of prices to be the main point of contention between "auction market theorists" and "disequilibrium theorists."

The overshooting model provides the necessary conditions for monetary policy to have short-run effects on relative prices of different sectors in the economy exhibiting differing degrees of price flexibility. The speed of adjustment of prices was shown to be a function of several parameters characterizing the economic system even in the most simplistic version of the model. Thus, some evidence is needed that would justify the assumption of price stickiness and/or differential price responses (across sectors) to changes in money. To be more specific, for the case of agriculture, some evidence is needed to justify characterizing the agricultural sector as the flexible price sector when compared to manufacturing and services sectors.

Although little doubt exists about the price flexibility of agricultural commodities, evidence (both theoretical and empirical) is needed to justify the assumption of price stickiness of other sectors in the economy. The costliness of continuous adjustment of prices seems to be the prevalent reason for price stickiness in the literature dealing with the microfoundations of macroeconomics.

Mussa (1981) recognizes the theoretical problems that arise in imposing rational expectation on models with sticky prices. He develops a price-adjustment rule that ". . . circumvents these theoretical difficulties and analyzes the essential economic characteristics of this rule" (p. 1021). The rule is derived from a microeconomic model in which there is an explicit cost in continuously changing prices, and thus it is optimal to adjust individual prices only at discrete intervals and by finite amounts. Prices change at such a frequency as to have the marginal gains associated with reducing disequilibrium, equal to the marginal costs of continuous price changing.

Also based on the "cost of adjustment" principle is the model by Rotenberg (1982). Like Mussa, he assumes that there are costs associated with changing prices, a fact that makes actual prices slowly respond to desired prices. He builds a dynamic model in which he incorporates a perceived cost of adjustment by firms, and he arrives at a form of the stickiness hypothesis amenable to empirical testing. Estimation of the theoretically derived price path satisfies all the relevant theoretical constraints and seems fairly robust to alternative specifications. The empirical results support the sticky-price hypothesis. A nested hypothesis of a "Walrasian adjustment" (instantaneous price adjustment to contemporaneous changes of money balances) is rejected by the data. As reasons for the sluggish price adjustment, he cites: (1) a small response of aggregate demand to changes in real-money balances and (2) high costs of changing prices perceived by firms. Another important implication comes from the strength that his results gain when food prices and fuel prices are removed from the price index (gross domestic product price deflator). This further supports the fix-price, flex-price separation and the characterization of agriculture as a flex-price sector.

On the empirical side, Gordon (1975) examined the Sargent-Wallace-Lucas proposition of instantaneous price response to money supply by regressing the quarterly percentage change in the nonfood price deflator on a distributed lag of money supply growth.

Although these results show a strong relationship between price changes and money (lagged coefficients sum up to 1.366 after 28 quarters), only 14 percent of that change is felt by the end of two years and only 35 percent within four years. Gordon's conclusion was that much of the inertia lies in the influence of unemployment on wages.

Bordo (1980) constructed a model in which he related price variability in different sectors to contract length. He concluded that sectors with longer contract lengths exhibit lower price variability. Using price variability as a proxy for price flexibility, he classified commodities as flexible price markets while commodity prices were found to respond more rapidly than prices of manufactures to monetary changes. His empirical results show significant differences between the price behavior of "auction" as compared to "customer" markets as classified by Okun.

Frankel and Hardouvelis (1985) found that, when a surprise occurs, nominal interest rates and commodity prices move in opposite directions. They argue that can only be explained using a fix-price, flex-price model. If the flex/flex specification were correct, either both rise (if the announcement causes the public to revise upward its expectations of future money growth) or else both fall (if the public revises downward its expectation of future money growth). The only hypothesis that explains the reactions in both the interest rate and commodity markets is that the increase in the nominal interest is also an increase in the real interest rate. This is presumably because the public anticipates that the Federal Reserve will reverse the recent fluctuations in money stock thus increasing interest rates and depressing the real prices of commodities. Lombra and Mehra, in examining the effects of monetary and fiscal policy on different prices, found a larger but slower effect of money on food prices along the marketing chain.

Stamoulis (1985, chapter 3) used the overshooting model to derive testable hypotheses concerning the relationship between the degree of overshooting of farm prices and the number of flexible price markets in the economy. These results support the predictions of the theoretical model in Section 2 and indicate that, as the number of flexible prices increases or, alternatively, as the weight of flexible prices in the general price index increases (α goes down), the degree of overshooting is reduced. This result agrees with the result of equation (12).

Using a combined macroeconomic-agricultural sector model of the U.S. economy, Rausser et. al. (1986) demonstrated that monetary policy could have strong short-run effects on relative prices of basic agricultural commodities (wheat, corn, and livestock products). Their results suggest strong real effects of monetary and fiscal policies on the farm sector of the U.S. economy. Finally, Frankel (1986) constructed a theoretical model in which he considered storable agricultural commodities as assets, linking their prices to the financial market by the basic arbitrage condition of stockholding (i.e., that the expected rate of change of commodity prices should not exceed the rate of interest). The model produces results similar to those of Section 2 (i.e., if some prices in the economy are characterized by stickiness and if real output is constant in the short run, then flexible prices overshoot their long-run equilibrium).

EMPIRICAL EVIDENCE

To test for differential effects of money growth on prices in different sectors, the following model was estimated using percentage changes:

$$\dot{p}_{it} = \alpha_0 + \delta \dot{p}_{i,t-1} + \sum_{j=0}^{K} \beta_j \dot{m}_{t-j} + \sum_{j=0}^{3} \gamma_j \dot{g}_{t-j} \qquad (26)$$

where:

$\dot{p}_{it}$ = the growth rate of price index i at time t

$\dot{m}_t$ = the growth rate of money at time t

$\dot{g}_t$ = the growth rate of real gross national product (GNP) at time t

The lag length, K, was chosen by maximizing $\bar{R}^2$ over the range of models (differing by the lag length on m) in which money neutrality could not be rejected. A maximum of 12 lags was tried. The index of prices received by farmers (IPRF) was chosen to represent flexible prices while the nonfood component of the consumer price index (CPINF) represented the sticky prices in the economy. Results were also obtained for the consumer price index for food (CPIF). Quarterly dummies were included in all models, and a lagged dependent variable was used to capture partial adjustment effects. The magnitude and significance of that variable could also serve as an indicator of price flexibility. To account for any possible short-run cyclical effects of output on the path of prices, real GNP growth variables were included in the equations.

Although one could use the above model to classify prices in terms of their flexibility, the focus was on establishing a basis for examining how monetary policy causes relative price changes which affect the farm sector. Thus, results obtained using the IPRF are distinguished from results obtained using the consumer price index for food and beverages (CPIF) which contains a sizable marketing component.

Regression results obtained using OLS are reported in Table 5.1. To test for the neutrality of money, the null hypothesis is that the sum of the coefficients of lagged money equals $(1-\delta)$ where δ is the coefficient of the lagged dependent variable. The test can be derived from equation (26) by observing that in the long-run:

$$\dot{p}_{i,t} = \dot{p}_{i,t-1}$$

and that:

$$\dot{m}_{t-j} = \dot{m}_{t-\ell},\ \dot{g}_{t-j} = \dot{g}_{t-\ell},\ \text{for all } j,\ell;$$

thus, the long-run effect of money on prices can be derived as:

$$\dot{p}_{1r}(1-\delta) = \sum_j \beta_j\, \dot{m}_{1r} + \sum_j \gamma_j\, \dot{g}_{1r}$$

and:

$$\frac{d\dot{p}_{1r}}{d\dot{m}_{1r}} = \frac{\sum_j \beta_j}{1-\delta}\ .$$

So:

$$H_0: \frac{\sum_j \beta_j}{1-\delta} = 1 \quad \text{or} \quad \sum_j \beta_j = 1 - \delta \tag{27}$$

Equation (27) thus becomes the null hypothesis for the neutrality test. On the basis of this test and the R^2 criterion, a model with only contemporaneous effects of money growth was chosen for the IPRF index while the results showed that, for the CPINF, a model with 10 lags on money growth was appropriate.

In analyzing the empirical results, several aspects relating to predictions of the theoretical model are of interest. Namely, tests were performed on the differential effects of money on the several price indices, the neutrality hypothesis [as expressed in (27)], the overshooting hypothesis $\frac{dp_t^{IPRF}}{dm_t} > 1$, and the differential speed of adjustment of various indices (as expressed by the coefficient of the lagged dependent variable).

TABLE 5.1
Regression Results for DIPRF and DCPINF

Variable	DIPRF	DCPINF
C	-4.877	-0.815
	(2.329)[a]	(0.720)
$\dot{p}_{t-1}$	0.113	0.692
	(0.147)	(0.147)
$\dot{g}_t$	0.267	0.051
	(0.721)	(0.135)
$\dot{g}_{t-1}$	0.190	0.068
	(0.694)	(0.125)
$\dot{g}_{t-2}$	-0.293	-0.025
	(0.691)	(0.120)
$\dot{g}_{t-3}$	1.046	0.043
	(0.662)	(0.124)
$\dot{m}_t$	1.568	0.104
	(0.953)	(0.189)
$\dot{m}_{t-1}$		-0.099
		(0.180)
$\dot{m}_{t-2}$		-0.077
		(0.167)
$\dot{m}_{t-3}$		-0.033
		(0.172)
$\dot{m}_{t-4}$		0.028
		(0.166)
$\dot{m}_{t-5}$		0.144
		(0.199)
$\dot{m}_{t-6}$		0.260
		(0.176)
$\dot{m}_{t-7}$		0.310
		(0.167)
$\dot{m}_{t-8}$		0.195
		(0.174)
$\dot{m}_{t-9}$		-0.170
		(0.196)
$\dot{m}_{t-10}$		0.076
		(0.186)

$\bar{R}^2$ DIPRF = 0.067 $\bar{R}^2$ DCPINF = 0.516

H_0 : $\delta + \beta_0 = 1$

$F_{1,33}$: 0.482, P > F = 0.491

H_0 : $\delta + \sum_{i=0}^{10} \beta_i = 1$

$F_{1,33}$: 1.031, P > F = 0.317

[a]Figures in parentheses are standard errors.

The instantaneous effect on flexible farm prices is estimated to be 1.57 which suggests overshooting. While the associated standard error is large, the point estimate and the small and insignificant partial adjustment effect are consistent with the price flexibility assumption. In contrast, for the CPINF, both the instantaneous effect and the coefficient and significance of the lagged dependent variable suggest slower adjustment. The sum of the coefficients on money growth and the lagged dependent variable coefficient is not significantly different from 1 by construction. Later lags on money are more significant than earlier ones.

The distinction between a rapid adjustment of farm prices as opposed to a slow one for nonfarm ones is consistent across lag lengths. In fact, a model for farm prices with 10 lags features a larger and more statistically significant instantaneous effect of money growth. To test hypotheses across equations, a second set of results was derived by jointly estimating the preferred models using Zellner's Seemingly Unrelated Regressions (SUR) technique. Note that the results for this approach shown in Table 5.2 are not substantially different from those presented in Table 5.1. The joint test for an equal instantaneous response of the two prices to the monetary shock cannot be rejected at the 10 percent significance level.

In the theoretical model of Section 2, overshooting of flex prices to a shock was defined in reference to the long-run equilibrium in which the whole system reaches equilibrium simultaneously (both flex and fixed prices). Since for the case of 10 lags on money growth, the sum of the coefficients does not significantly differ from 1 for either price index and given the selection criterion for lag length, the parameters of the two models were jointly estimated using SUR assuming a lag length of 10 for both. Results are shown in Table 5.3. The magnitude and significance of the instantaneous effect of money on the IPRF are as suggested by theory. Both the magnitude of the coefficients and t-statistics suggest a strong and significant reaction of IPRF to changes in money growth. An F-test at the 5 percent level revealed that the instantaneous responses of the two prices differ significantly, a result that supports the assumption in the theoretical model. A joint test that the sum of the coefficients on money in each of the models is equal to 1 cannot be rejected at the 5 percent significance level. Results of the tests are also presented in Table 5.3.

Table 5.4 includes the results for the best-fitting model for the CPIF. The behavior of this regression is closer to that of the

TABLE 5.2
Regression Results for DIPRF and DCPINF

Variable	DIPRF	DCPINF
C	-4.954	-0.883
	(2.329)[a]	(0.716)
$\dot{p}_{t-1}$	0.144	0.671
	(0.146)	(0.146)
$\dot{g}_t$	0.238	0.059
	(0.721)	(0.135)
$\dot{g}_{t-1}$	0.177	0.083
	(0.695)	(0.125)
$\dot{g}_{t-2}$	-0.295	-0.027
	(0.691)	(0.120)
$\dot{g}_{t-3}$	1.048	0.045
	(0.662)	(0.124)
$\dot{m}_t$	1.591	0.109
	(0.953)	(0.189)
$\dot{m}_{t-1}$		-0.137
		(0.179)
$\dot{m}_{t-2}$		-0.082
		(0.166)
$\dot{m}_{t-3}$		-0.045
		(0.171)
$\dot{m}_{t-4}$		0.053
		(0.165)
$\dot{m}_{t-5}$		0.122
		(0.198)
$\dot{m}_{t-6}$		0.296
		(0.174)
$\dot{m}_{t-7}$		0.327
		(0.165)
$\dot{m}_{t-8}$		0.220
		(0.172)
$\dot{m}_{t-9}$		-0.175
		(0.194)
$\dot{m}_{t-10}$		0.120
		(0.184)

H_0 : $DM_{IPRF} = DM_{CPINF}$

$F_{1,76}$: 2.016, $P > F = 0.159$

[a] Figures in parentheses are standard errors.

TABLE 5.3
Regression Results for DIPRF and DCPINF

Variable	DIPRF	DCPINF
C	-8.469	-0.820
	(5.261)[a]	(0.720)
$\dot{P}_{t-1}$	0.212	0.667
	(0.169)	(0.145)
$\dot{g}_t$	0.474	0.054
	(1.012)	(0.135)
$\dot{g}_{t-1}$	1.066	0.068
	(0.904)	(0.125)
$\dot{g}_{t-2}$	-0.452	-0.025
	(0.873)	(0.120)
$\dot{g}_{t-3}$	1.403	0.040
	(0.891)	(0.124)
$\dot{m}_t$	2.778	0.090
	(1.255)	(0.188)
$\dot{m}_{t-1}$	-2.192	-0.103
	(1.305)	(0.180)
$\dot{m}_{t-2}$	-0.060	-0.079
	(1.234)	(0.167)
$\dot{m}_{t-3}$	-0.902	-0.030
	(1.237)	(0.172)
$\dot{m}_{t-4}$	1.591	0.026
	(1.193)	(0.166)
$\dot{m}_{t-5}$	-1.917	0.151
	(1.500)	(0.200)
$\dot{m}_{t-6}$	1.828	0.267
	(1.226)	(0.176)
$\dot{m}_{t-7}$	0.554	0.318
	(1.164)	(0.167)
$\dot{m}_{t-8}$	0.847	0.206
	(1.173)	(0.174)
$\dot{m}_{t-9}$	-1.268	-0.153
	(1.238)	(0.195)
$\dot{m}_{t-10}$	2.501	0.080
	(1.330)	(0.186)

H_0 : $DM_{IPRF} = DM_{CPINF}$

$F_{1,66}$: 4.330, $P > F = 0.041$

H_0 : $\frac{1}{1-\delta} * \sum_{i=0}^{10} \beta_i^{DIPRF} = 1, \frac{1}{1-\delta} * \sum_{i=0}^{10} \beta_i^{DCPINF} = 1$

$F_{2,66}$: 1.1998, $P > F = 0.308$

[a]Figures in parentheses are standard errors.

TABLE 5.4
Regression Results for DCPIF

Variable	DCPIF
C	-0.981 (1.401)[a]
$\dot{p}_{t-1}$	0.532 (0.158)
$\dot{g}_t$	0.100 (0.199)
$\dot{g}_{t-1}$	0.459 (0.196)
$\dot{g}_{t-2}$	-0.166 (0.213)
$\dot{g}_{t-3}$	0.072 (0.197)
$\dot{m}_t$	0.066 (0.281)
$\dot{m}_{t-1}$	-0.305 (0.262)
$\dot{m}_{t-2}$	-0.375 (0.276)
$\dot{m}_{t-3}$	-0.028 (0.312)
$\dot{m}_{t-4}$	0.607 (0.245)
$\dot{m}_{t-5}$	-0.005 (0.317)
$\dot{m}_{t-6}$	0.227 (0.269)
$\dot{m}_{t-7}$	0.195 (0.267)
$\dot{m}_{t-8}$	0.396 (0.267)
$\dot{m}_{t-9}$	-0.102 (0.274)
$\dot{m}_{t-10}$	0.193 (0.273)
$\dot{m}_{t-11}$	0.447 (0.283)
$\dot{m}_{t-12}$	-0.396 (0.310)

$H_0 \quad : \frac{1}{1-\delta} * \sum_{i=0}^{12} \beta_i = 1$

$F_{1,31}$: 0.344, $P > F = 0.562$

[a]Figures in parentheses are standard errors.

CPINF than it is to the IPRF. Lombra and Mehra (1983) found a larger but slower effect of money growth on food prices the further along the marketing chain, so this is consistent with their results. Table 5.4 shows that the fourth lag on money growth is the most important of the 12 lags and that the neutrality hypothesis is rejected for shorter lag lengths. Also indicating the stickiness of the CPIF is the large and significant coefficient on the lagged dependent variable.

Finally, Table 5.5 presents results from the joint estimation (using SUR) of the three equations. For reasons explained above, a 10th order lag was specified for all the equations. No substantial change in the results and the conclusions derived in the previous models occurs although the significance of the instantaneous effect of money on the flexible price index is reduced. Pairwise F-tests for the equality of instantaneous response coefficients across price indices are also shown in Table 5.5. The hypothesis of equal instantaneous responses between IPRF and CPINF is reflected at the 10 percent significance level. The equality of the IPRF and CPIF responses is rejected at the 5 percent level while the hypothesis of equal instantaneous responses of CPIF and CPINF cannot be rejected.

The extent to which flexible prices overshoot their long-run equilibrium following a monetary shock depends on several parameters, as shown in (12). It is increasing in α (the share of fixed prices in the price index) and decreasing in λ (the semi-elasticity of money demand with respect to the interest rate). Thus, changes in the economy affecting these parameters can be expected to cause changes in the relationship between price changes and money growth.

It was hypothesized that changes in monetary policy such as the shift to floating exchange rates and the targeting of reserves in October of 1979 could show up as shifts in these parameters. To the extent that moving to floating exchange rates reduced the degree of insulation of some sectors from world prices, α might have fallen in the early 1970s. Similarly, apparent increases in λ over time that were found when a simple money demand equation was estimated suggested that had risen with interest rates after 1979.

Chow tests were performed to see if expected changes in price flexibility followed. In the preferred model for farm prices, no statistically significant change in the coefficient of instantaneous effect of money growth was found, but the sign accorded with

TABLE 5.5
Regression Results for DIPRF, DCPINF, and DCPIF

Variable	DIPRF	DCPINF	DCPIF
C	-9.633	-0.824	-1.044
	(5.232)[a]	(0.720)	(1.091)
$\dot{p}_{t-1}$	0.000	0.643	0.459
	(0.136)	(0.145)	(0.129)
$\dot{g}_t$	0.847	0.056	0.113
	(0.997)	(0.135)	(0.205)
$\dot{g}_{t-1}$	1.181	0.068	0.372
	(0.903)	(0.125)	(0.189)
$\dot{g}_{t-2}$	-0.287	-0.025	-0.114
	(0.869)	(0.120)	(0.189)
$\dot{g}_{t-3}$	1.235	0.036	0.118
	(0.888)	(0.124)	(0.185)
$\dot{m}_t$	2.343	0.076	0.065
	(1.298)	(0.188)	(0.265)
$\dot{m}_{t-1}$	-1.902	-0.107	-0.279
	(1.298)	(0.180)	(0.270)
$\dot{m}_{t-2}$	-0.437	-0.082	-0.224
	(1.222)	(0.167)	(0.253)
$\dot{m}_{t-3}$	-0.801	-0.028	-0.180
	(1.236)	(0.172)	(0.258)
$\dot{m}_{t-4}$	1.547	0.025	0.513
	(1.193)	(0.166)	(0.250)
$\dot{m}_{t-5}$	-1.215	0.159	0.032
	(1.463)	(0.199)	(0.315)
$\dot{m}_{t-6}$	1.843	0.274	0.417
	(1.226)	(0.175)	(0.264)
$\dot{m}_{t-7}$	0.736	0.325	0.003
	(1.161)	(0.167)	(0.248)
$\dot{m}_{t-8}$	1.023	0.216	0.530
	(1.170)	(0.173)	(0.244)
$\dot{m}_{t-9}$	-0.980	-0.137	-0.088
	(1.230)	(0.195)	(0.273)
$\dot{m}_{t-10}$	2.398	0.083	0.267
	(1.329)	(0.186)	(0.278)

H_0 : $DM_{IPRF} = DM_{CPINF}$
$F_{1,99}$: 3.208, $P > F = 0.076$
H_0 : $DM_{IPRF} = DM_{CPIF}$
$F_{1,99}$: 4.515, $P > F = 0.036$
H_0 : $DM_{CPINF} = DM_{CPIF}$
$F_{1,99}$: 0.001, $P > F = 0.973$

[a]Figures in parentheses are standard errors.

expectation indicating a positive relationship. That is, the effect of money on prices appears to have increased slightly after 1973 and decreased in 1979 and after.

SUMMARY, CONCLUSIONS, AND POLICY IMPLICATIONS

A theoretical model was presented to show that, when prices in various sectors differ in their speed of adjustment to different shocks, money growth changes could cause flexible prices to overshoot their long-run equilibrium. However, the importance to agricultural economists is not the overshooting result per se but the possibility that relative prices of farm products can be affected by monetary policy.

Empirical results for the United States reported here lend support to the proposition that monetary changes cause short-run changes in relative farm prices although the overshooting hypothesis could not be confirmed. This is not surprising given that farm prices differ between commodities as to the degree of flexibility. Several prices entering the index of prices received by farmers exhibit downward inflexibility because of government price supports. The results obtained by Frankel and Hardouvelis (1985) for specific commodities would support the observation that the price index data used here to represent flexible prices probably understate the distinction between those commodities that actually have perfect price flexibility and those of the other sectors of the economy.

The main empirical result (i.e., that changes in relative prices follow a change in money growth) seems to be well supported by other evidence found in the literature. In addition to the reduced form regressions presented here, more structured single-equation models (Stamoulis, 1985) seem to support this hypothesis. In addition, results obtained using Australian data (Chalfant et. al., 1986) also support the main assumptions of the theoretical model.

Changes in relative prices following monetary changes may partly explain the drastically different environments facing the farm sector during the early 1970s vs. 1980s. According to the model, the "easy money" policies of the early 1970s should have contributed to the rise in relative prices of commodities. On the other hand the tight monetary policies of the early 1980s should have contributed to the decline in relative farm prices.

According to Rausser et. al. (1986), "A fair characterization of the monetary and fiscal policies of the early 1970s and the early

1980s is that the first period represented a subsidy period for agriculture while the latter regime taxed the sector!" In other words, stickiness in other sectors caused agriculture to prosper in the early 1970s and decline in the early 1980s.

The relevant question for policymakers is whether or not there is need for policy action when monetary policy creates an adverse environment for the farm sector. If monetary policy causes problems to agriculture for reasons outside the sector (i.e., stickiness in other sectors), is agricultural policy legitimized?[3] In other words, can we consider farm policy as correcting the externality that monetary policy imposes on the farm sector due to stickiness in other sectors?

There is no clear-cut answer to this question since the answer depends on the objectives of agricultural policy itself. If monetary policy is long-run neutral and the short-run effects will eventually be reversed, then, conceivably, no policy action needs to be taken.

But there are possible exceptions. Suppose farmers have myopic expectations and build capacity on the basis of short-run movements in relative prices and agriculture is characterized by asset fixity. If both hold, conceivably farmers would build excess capacity when relative farm prices are high, following a series of monetary shocks. Then, when the initial effects of money start reversing themselves, resources in the farm sector cannot adjust because of asset fixity. The result is overcapacity, excess supply and--in the absence of policy action--further price declines.

In such a case, a policy combination of income enhancement and supply control may be needed. The danger is that a scheme of price supports associated with ineffective supply control will inhibit adjustment of resources to their long-run equilibrium.

It is in the above sense that one could make the claim that the policies of the 1970s set the stage for the events of the early 1980s. Although the evidence is overwhelming that monetary policy has a strong short-run effect on the agricultural sector, if money is neutral in the long-run, then agricultural sector policies are likely to have a more significant influence on the farm sector.

It is worth noticing that there is an asymmetry in the types of public policy actions triggered by alternative macroeconomic environments. In periods of favorable macroeconomic environments, no action is taken toward reducing relative farm prices. Then in periods of macroeconomic conditions that "tax" the farm sector, the most costly farm programs are implemented.

A possible course of action for agricultural policy is a flexible policy scheme as proposed by Just and Rausser (1984). According to that concept, farm policy should be designed conditional on macroeconomic conditions facing agriculture. The basic concept in those policies is the flexible character of both storage and target price policies. Conceptually, a flexible policy should impose a self-regulating tax during "subsidy type" macroeconomic environments while a self-regulating subsidy will be imposed during a "tax type" macroeconomic environment.

NOTES

1. Note that undershooting does not mean that relative prices in the economy do not change following a monetary shock. It simply means that they deviate from the long-run equilibrium by less than in the case of overshooting. For a hypothetical change in m of, for example, 10 percent overshooting implies that e and p_A will change more than 10 percent while undershooting implies that they change by less than 10 percent.

2. For a derivation of alternative overshooting coefficients and adjustment paths under different assumptions about output adjustment and expectations, see Rausser, Nishiyama, and Stamoulis (1986).

3. We assume here that no one is concerned about implementation of agricultural policy to deal with periods that are "too good" for the farm sector.

REFERENCES

Batten, D. S., and M. T. Belongia, "Money Growth, Exchange Rates and Agricultural Trade," Federal Reserve Bank of St. Louis, unpublished manuscript, 1984.

Bordo, M.D., "The Effects of Monetary Change on Relative Commodity Prices and the Role of Long Term Contracts," Journal of Political Economy. 88(1980): 1088-1094.

Chalfant, J. A., H. Alan Love, G. C. Rausser, and K. G. Stamoulis, "The Effects of Monetary Policy on U.S. Agriculture." Paper presented at the 30th annual meetings of the Australian Agricultural Economics Society, Canberra, 1986.

Chambers, R. G., and R. E. Just, "An Investigation of the Effects of Monetary Factors on Agriculture," Journal of Monetary Economics. (1982): 235-247.

Dornbusch, R., "Expectations and Exchange Rate Dynamics," Journal of Political Economy. 84(1976): 1161-1176.

Driskill, R. A., "Exchange Rate Dynamics, Portfolio Balance and Relative Prices," American Economic Review. 70(1980): 776-783.

Engel, C.M., and R.P. Flood, "Exchange Rate Dynamics with Wealth Effects: Some Theoretical Ambiguities," Journal of Money, Credit, and Banking. 17(1985): 312-327.

Frankel, J. A., "Expectations and Commodity Price Dynamics: The Overshooting Model," American Journal of Agricultural Economics. 68(1986): 344-350.

Frankel, J. A., and G. Hardouvelis, "Commodity Prices, Monetary Surprises, and Fed Credibility," Journal of Money, Credit, and Banking. 17(1985): 425-438.

Gordon, R. J., "The Impact of Aggregate Demand on Prices." Brookings Papers on Economic Activity, 1975.

Gordon, R. T., "Price Inertia and Policy Ineffectiveness in the United States; 1890-1980," Journal of Political Economy. 90(1982): 1087-1117.

Just, R. E., and G. C. Rausser, "Uncertain Economic Environments and Conditional Policies," Alternative Agricultural and Food Policies and the 1985 Farm Bill. G. C. Rausser and K. R. Farrell. (eds.) Giannini Foundation of Agricultural Economics. Division of Agricultural and Natural Resources. University of California, Berkeley, 1984.

Lombra, R. E., and Y. P. Mehra, "Aggregate Demand, Food Prices, and the Underlying Rate of Inflation," Journal of Macroeconomics. 5(1983): 383-398.

Mussa, M., "Sticky Prices and Disequilibrium Adjustment in a Rational Model of the Inflationary Process," American Economic Review. 71(1981): 1020-1027.

Obstfeld, M., "Overshooting Agricultural Commodity Markets and Public Policy: Discussion," American Journal of Agricultural Economics. 68(1986), 420-421.

Okun, A. M., Prices and Quantities. Washington, D.C., The Brookings Institution, 1981.

Rausser, G.C., J.A. Chalfant, H.A. Love, and K.G. Stamoulis, "Macroeconomic Linkages, Taxes, and Subsidies in the Agricultural Sector," American Journal of Agricultural Economics. 68(1986): 399-417.

Rausser, G.C., Y. Nishiyama, and K.G. Stamoulis, "Macroeconomics, Overshooting, and the U.S. Agricultural Sector." Working Paper No. 410. University of California, Berkeley, Department of Agricultural and Resource Economics, 1986.

Rotenberg, J. J., "Sticky Prices in the United States," Journal of Political Economy. 90(1982): 1187-1211.

Stamoulis, K. G., "The Effects of Monetary Policy on United States Agriculture. A Fix-Price, Flex-Price Approach," Unpublished Ph.D. Dissertation, Department of Agricultural and Resource Economics, University of California, Berkeley, December, 1985.

6
Exchange Rates and Macroeconomic Externalities in Agriculture

Richard E. Just

International agricultural trade has increasingly taken on macroeconomic dimensions over the last 15 years. These changes began in the early 1970s with the shift away from the Bretton Woods system of fixed exchange rates. The ensuing exchange rate volatility clearly illustrated the importance of macroeconomic policy to agriculture through its effect on exchange rates (Schuh, 1974). Some empirical studies have found exchange rates to be the dominant force affecting major agricultural exports and prices (Chambers and Just, 1981; Longmire and Morey, 1983). Partly because of this dominant role of exchange rates in major agricultural markets, several studies have found that the effects of macroeconomic policies on agriculture can more than offset agricultural policies in terms of the relative price signals guiding producers and consumers (Rausser, 1986; Valdes, 1986).

Another event that has increased the macroeconomic dimensions of agricultural trade is the development of the international monetary system associated with the development of the Eurocurrency market and the ensuing flood of petro dollars. With the emergence of this international capital market, highly fluid international financial flows have grown to more than an order of magnitude larger than international trade. This development, in conjunction with adoption of a system of floating exchange rates among major currencies, has led to worldwide monetary instability and highly volatile agricultural trade conditions that are critically dependent on macroeconomic policy. These conditions, after a

Richard E. Just, Professor, Department of Agricultural and Resource Economics, University of Maryland, College Park, MD.

decade of monetary instability, have led some to assert that monetary and fiscal stability is more important in stabilizing agricultural markets than sector-specific instruments such as grain reserves and suggest that a disproportionate share of the burden of adjustment to macroeconomic fluctuations is imposed on trade sectors of the economy (Schuh and Orden, 1985). These arguments suggest that domestic agricultural and domestic macroeconomic policy are inseparable in the sense that the former cannot be studied without considering its interaction with the latter.

Furthermore, in the same way that the monetary instability of the last decade has revealed that domestic macroeconomic and domestic agricultural policies are inseparable, it has also revealed that domestic and foreign policies are inseparable. Evidence of the importance of foreign macroeconomic policy to agriculture is clear from a developing country perspective because of the increased portion of adjustment in world agricultural trade that has involved middle income developing countries in recent years. For example, during the late 1970s when the United States was pursuing a liberal monetary policy that led to devaluation of the dollar, these countries accounted for almost half of the increased value of U.S. agricultural exports. They also accounted for almost half of the decline in 1982 when the United States sharply tightened its monetary policy and the dollar appreciated rapidly.

Furthermore, while following restrictive monetary policy, the United States pursued an expansionary fiscal policy while significant European trading partners adopted restrictive fiscal policy and more expansionary monetary policy. The result was to create a significant flow of capital towards the United States. As a result, the dollar appreciation hurt the position of U.S. agricultural exports more than if either the U.S. or European macroeconomic policies had remained unchanged.

While these various stories have been told and retold many times, they are associated with a growing problem of macroeconomic externalities in agriculture. The purpose of this paper is to examine the extent to which macroeconomic effects on agriculture should be classified as external effects and, thus, considered as problems of market failure. The paper argues that macroeconomic considerations should differ depending on the extent to which external effects are present. Some of these externalities occur because traditional market failures play major roles in agricultural markets while others are due to myopic views of the world by public and/or private decision makers. Some of the considerations are specific to agriculture while others apply to other industries as well.

Consideration of macroeconomic externalities is not new with this paper. For example, Friedman (1959a), Kolm (1972), Klein (1974), and Niehans (1978) have alluded to confidence externalities associated with money. Hall (1981) has suggested that "maintenance of a stable price level carries significant positive externalities" (p. 1). The work of Friedman (1959b; 1969), Chetty (1969), and Fischer (1974) suggests the possibility of transaction cost externalities associated with different individuals using the same currency. Vaubel (1984) extends these considerations to evaluate externalities associated with fixing exchange rates. While he argues that these externalities tend to be pecuniary and thus not Pareto relevant, he does so only in the context of an otherwise perfect economy.

While many of the arguments in this paper suggest that macroeconomic externalities can be reduced or eliminated through exchange rate stability, the paper should not be construed as an argument for fixing exchange rates. The intent is simply to point out some adverse and often overlooked effects of exchange rate and macroeconomic instability that should be balanced against other perhaps beneficial effects in determining an appropriate level of stability, and to explore possibilities for reducing those adverse effects. While this paper focuses on the effects in agriculture, the possibility of similar effects in other sectors suggests that these considerations may be important at the aggregate level as well.

The paper begins with a general discussion of what constitutes a macroeconomic externality and of evidence that macroeconomic externalities are of increasing importance. Most of the remainder of the paper discusses specific kinds of macroeconomic externalities of importance in agriculture. Finally, prospects for reducing or eliminating macroeconomic externalities are considered.

MACROECONOMIC EXTERNALITIES

An externality is traditionally defined as an effect of one decision maker on another that is not fully reflected through the market place. An externality occurs when a decision maker does not take into account the true marginal social impact of his decision. Traditionally, externalities are viewed as occurring among private decision makers. In this paper, public decision makers are also involved through their role as providers of public goods such as money, exchange rate stability, and consistency of monetary and fiscal policies. Thus, the crucial question is whether public decision makers take into account the true marginal social impact of their decisions.

For purposes of discussion, public decision makers in this paper are assumed to pursue efficiency oriented objectives but with limited information. (More general objectives relating to distribution could also be incorporated into the discussion with the same resulting principles but some of the arguments become unnecessarily complex.) Public policy externalities occur either because of lack of public infrastructure for determining and implementing policies or because of lack of information. Public policy externalities occur with limited information because public decision makers are unable to take into account the full marginal social impact of their decisions.

Public goods may be provided at socially inappropriate levels for a variety of reasons. First, errors may be caused by traditional forms of market failure. For example, a government may be viewed as a monopoly producer of a currency. However, this would not necessarily imply a public policy externality even though a public good is normally a special case of externality. When public policy makers have complete information and pursue efficiency objectives, the usual public good externality is internalized. Alternatively, Vaubel (1984) argues that the externalities and monopoly problems associated with money are due to restricting private competition in its provision. This paper, however, is cast in a (more realistic?) context taking existing imperfections as given. This gives rise to a second reason why public goods may be provided at inappropriate levels. Namely, errors occur if providers of public goods do not account for other market failures in deciding how much to provide. The problem of providing money, for example, may differ drastically depending on capital market imperfections. Internalizing these externalities in a public policy framework depends crucially on information about how these market failures are affected by policy. Third, errors may occur because providers of public goods have poor expectations or make incorrect judgments. The history of macroeconomic policy of the past two decades is replete with such examples. Again, the problem is information based. Fourth, an institution that is willing and capable of providing a needed public good may not exist. For example, a government may not exist or have the ability to provide certain kinds of public goods.

In the context of international trade, the provision of public goods must be considered in an international framework. One dilemma is how to provide international public goods in appropriate amounts in absence of world government. Bryant (1980), who is one of the first to discuss international public goods, argues that supranational organizations will tend to emerge through international cooperation to provide desirable levels of international public goods. In this context, he expects the International Monetary Fund (IMF), for example, to provide appropriate leadership for world cooperation in monetary and fiscal policy (p. 481). Given the events of the last fifteen years, however, these possibilities appear bleak. The IMF has been given neither the resources nor the power to act as a central banker for the world. Also, its special drawing right (SDR) standard has tended to give way to the U.S. dollar as an international monetary standard since the collapse of the Bretton Woods agreement when the two diverged.

In an interesting presidential address to the American Economic Association, Kindleberger (1986) argues that such idealistic international cooperation is unlikely to occur citing the interwar record of the League of Nations as an example. He argues that international public goods, such as international money, stable exchange rates, and consistent macroeconomic policies are likely to be provided by the political scientists' so-called "hegemon", if at all. A hegemon is a leading world power which may be willing to bear an undue share of the short-run costs of an international public good in return for long-run gains possibly of a nonpecuniary nature. Alternatively, a hegemonic world power may simply set in motion habits of international cooperation around which expectations of international actors converge in given issue areas (Krasner, 1983). For example, the British hegemony of the nineteenth century fostered free trade and the gold standard while the subsequent American hegemony fostered various institutions and agreements such as Bretton Woods. Kindleberger argues, however, that the American hegemony is in decline and that beginning about 1971 the United States "lost the appetite for providing international economic public goods" (p. 9).

This loss of international public goods suggests increased adverse effects of externalities (because public goods are special cases of externalities) unless they are internalized through such means as policy. The collapse of the Bretton Woods agreement, for example, has left countries free to try to export their domestic problems abroad through adjustment of macroeconomic policy. (With fixed exchange rates, countries were expected to manage macroeconomic phenomena relative to external accounts, unemployment, and inflation through domestic monetary and fiscal policy as opposed to currency devaluation.) As a result, some of the major shocks to the international economy such as the oil price shocks of 1973 and 1978 led to severe exchange rate instability and large macroeconomic policy adjustments in efforts to contain inflation, unemployment, and trade deficits. These adjustments led to further shocks such as the interest rate shock which occurred around 1980, the debt repayment crisis signaled by Mexican problems in August 1982, and the ensuing decline in international credit availability in 1983 and 1984 (see Khan, 1986, for a review of these events). These major shocks and the ensuing macroeconomic policy adjustments led to sharp increases in exchange rate and monetary instability which, in turn, led to increased severity of a variety of macroeconomic externalities and market failures.

IMPERFECT CAPITAL MARKETS

One of the most common examples of market failure in both developed and developing agriculture is the market for credit. Some farmers are unable to obtain credit in the amounts they need if at all. The market does not clear at the market interest rate. While this is a traditional form of market failure, it gives rise to a macroeconomic policy externality if appropriate information is not available on its response to macroeconomic policies.

Access to credit often depends on the relationship of interest rates (or payments) to asset values. Asset values in agriculture depend heavily on grain prices and export demand which, in turn, depend on exchange rates. More importantly, asset values vary greatly between land owners and renters. If macroeconomic policies raise interest rates relative to asset values, then credit market imperfections apparently increase as more farmers lose access to credit. Public decision makers who alter, say, the supply of money based on interest, inflation, unemployment, and exchange rates, etc., may be making decisions that affect other agents in ways not fully reflected in their information base (not fully reflected in the macroeconomic variables on which they act). If so, they are not accounting for the true marginal social impact of their decisions so a macroeconomic externality occurs.

Recent events in U.S. agriculture seem to offer an example. First, sharp rises in interest rates occurred in 1979-1981 as a result of macroeconomic policy. Subsequently, the rising value of the dollar caused a decline in agricultural export demand and agricultural land prices. Both of these effects caused interest rates to rise relative to asset prices. As a result, U.S. agriculture incurred its largest debt crisis since the Great Depression. The large number of farm foreclosures and the declining access to credit for farmers that ensued is evidence of the resulting increase in credit market imperfections that occurred as externalities of the associated macroeconomic policies.

Foreclosures present a serious problem in policy formulation because their costs are not reflected by marginal behavior. Costs of restructuring assets and debt or of retraining for other occupations are fixed costs that tend not to be reflected in economic markets even though they may be caused by marginal macroeconomic policy adjustments.

One might also note that positive consequences could result from extreme macroeconomic instability in this case. That is, macroeconomic instability could increase either the private or social cost of these imperfections to the point of inducing adjustments or formation of policies or institutions to reduce or eliminate them. If so, macroeconomic policy instability could have some fruitful effects which may or may not be anticipated.

One such result may be evident in the recent reformulation of lending practices in the Farm Credit System. Prior to the farm debt crisis of 1982, farm credit qualification in Federal Land Banks and Production Credit Associations depended largely on wealth and equity considerations whereas recent practices depend much more on a demonstrated ability to repay through cash flow. This step may reduce capital market imperfects facing agriculture. On the other hand, many small commercial banks have suffered such disastrous consequences that they are unwilling to lend to farmers on any terms.

RISK AVERSION

Risk aversion can also lead to macroeconomic externalities. With the maturing of the international capital market and the increased volatility of world markets in the last decade, the focus on risk in agriculture has shifted from internal sources associated with domestic production to external sources associated with world prices and exchange rates. The shift to flexible exchange rates has caused a dramatic increase in risk faced by agricultural concerns in world markets. For example, several studies have found that exchange rates now explain more grain price variation than any other factor (Just, 1984b). As a result, price risk has become the most important intermediate-run risk farmers face.

Friedman (1953) pointed out long ago that the variability of flexible exchange rates should generally reflect underlying instabilities in economic conditions. However, since the move to flexible exchange rates, the most important economic conditions have turned out to be the (artificial?) macroeconomic policies of major industrial countries rather than other natural economic phenomena (Willet, 1981). This fact, together with the many empirical studies that demonstrate the importance of risk aversion in farmers' behavior, implies that macroeconomic policy instability imposes serious losses on agriculture. The occurrence of these losses does not necessarily imply a macroeconomic externality. A shifting of income distribution away from agriculture could be consistent with pursuing efficiency objectives. The crucial considerations are whether risk causes externalities and market failures and whether these effects are taken into account in macroeconomic policy formulation.

Brainard and Cooper (1968) have argued that risks in important export sectors tend to be passed on to other sectors of the economy through their effects on trade conditions and that this constitutes an externality to the extent that other sectors are affected adversely by risk. If, however, the risks in important trade sectors are the result of macroeconomic policies, then all of these losses associated with risk aversion are macroeconomic externalities. Arrow and Lind (1970) have argued that governments can be risk-neutral decision makers in certain circumstances. However, this is only true if the government absorbs the risks of its decisions. When risks are imposed on risk averse private concerns through policy adjustments, these risks must be taken into account unless contingency markets are available through which these risks can be transferred to risk neutral private concerns. Of course, such contingency markets are lacking for most agricultural inputs and assets and may be inefficient even for major agricultural outputs.

COSTS OF ADJUSTMENT

Costs of adjustment can cause both behavior and welfare effects similar to risk aversion (Just, 1975). Farmers often incur costs of adjustment with major market swings. For example, when the United States was pursuing expansionary monetary policy and restrictive fiscal policy in the early 1970s, the low value of the dollar induced record export demand. As a result, farmers began to cultivate previously sodded land and invested in more machinery to accommodate increased cultivation. However, the sharp turn-around in macroeconomic policy less than a decade later, and the associated commodity price declines, left farmers over-invested in machinery and in need of resodding idle eroding lands.

To the extent that these commodity swings were caused by macroeconomic policy, the associated costs of adjustment may be macroeconomic externalities. That is, if the first set of macroeconomic policies was pursued without adequate regard for the costs of adjustment incurred both in response to these policies and to later macroeconomic policies, possibly made necessary by the first, then the associated decision makers were not recognizing the true marginal social cost of their decisions.

The history of macroeconomic policy over the past fifteen years suggests that the adverse effects of adjustment cost and risk aversion have received little attention. Policy makers, perhaps because of shortsighted views related to upcoming elections, have been willing to make sharp revisions in macroeconomic policies to attain short term goals with seemingly little regard for the long term effects. For example, tight monetary and expansionary fiscal or expansionary monetary and tight fiscal policy are not compatible in the long run but are often pursued in the short run to postpone certain types of adjustment.

Sacrificing long run benefits for short run gains may be consistent with pursuing efficiency objectives with a high social discount rate. However, adopting these policy inconsistencies from time to time imposes extreme instability and burdens of adjustment on trade sectors of the economy through exchange rate and interest rate adjustment which may not be evident in the information base accessible to policy makers. The increased instability, for example, causes increased reluctance to invest in future productivity capacity. Some of the current devaluation of agricultural assets, in addition to the depressing effects of current macroeconomic policies, may be due to aversion to the price risks caused by the recent history of revisions in macroeconomic policies or to the transactions costs that are incurred to restructure assets and debt in response to macroeconomic policies.

Kormendi and Meguire (1984, 1985) and Fry and Lilien (1986) have recently produced evidence that this effect is important at the aggregate level and tends in the long run to defeat some of the usual short-run goals of macroeconomic policy adjustment. For example, Fry and Lilien find that the variance of money growth shocks has a significant negative effect on both medium term and long term GDP growth. When these social impacts of macroeconomic policy instability are not fully recognized in policy formulation, macroeconomic externalities occur.

INDUCED SECTOR POLICY UNCERTAINTY

While the risk aversion and cost of adjustment effects discussed thus far are due to price uncertainty which is indirectly a result of macroeconomic policy uncertainty, agricultural policy uncertainty is also a major concern for farmers. When macroeconomic policies change, the effects of agricultural policy instruments can differ drastically.

For example, during the monetary expansion and fiscal conservation of the early 1970s, U.S. agricultural price supports became ineffective as a result of the declining value of the dollar and the associated booming export demand; on the other hand, the tight monetary policy and expansionary fiscal policy of the early 1980s caused price supports to be far above free market prices. This led to uncertainty on the part of farmers about whether and to what extent price supports would be reduced as evidenced by the heavy debate surrounding the 1985 farm bill (see Just and Rausser, 1984, for further discussion). Since these considerations add to the price uncertainty faced by farmers, they further exacerbate the adverse effects of risk aversion and costs of adjustment and thus contribute further to the externalities of macroeconomic policy.

OVERSHOOTING IN AGRICULTURE

Increasingly, agricultural commodity markets have come to be viewed as more flexible than those for many other sectors of the economy. This has occurred partially as the result of the explosion of trading in commodity exchanges during the 1970s. The apparent differences in price flexibility among markets has led to a dualistic view of the economy in which agricultural commodities are viewed as traded in flex-price markets while many other commodities are traded in fixed-price markets (Dornbusch, 1976; Mussa, 1982; Phelps and Taylor, 1977). Exchange rates can fall into either of these classifications depending on the exchange rate and macroeconomic policy regime.

Flex-price markets clear in the short run by price adjustments whereas fixed-price markets clear in the short run by quantity adjustments. The result of this dichotomy of adjustments is that unanticipated monetary disturbances affect relative commodity prices in the short run even though long run effects are neutral. For example, Cavallo is an analysis of exchange rate impacts on agriculture in Argentina noted that these conditions cause the effects of monetary and exchange rate policies to have short run real effects that differ from the long run where the law of one price becomes effective. As a result, the burden of monetary instability that is otherwise placed on the agricultural sector by virtue of its importance in trade is further exacerbated. This phenomenon has also been shown empirically to cause price overshooting in U.S. agricultural markets (by Stamoulis, Chalfant, and Rausser, 1985).

The underlying reasons for these differences in price flexibilities are not well researched. One reason could be a difference in costs of adjustment among sectors in which case the costs of adjustment may be allocated equitably. Higher perishability and long production lags may also explain some of the higher price variation in agriculture. Even in these cases, however, social loss clearly occurs as a result of macroeconomic instability with overshooting if a flex price temporarily drops low enough to induce disinvestment beyond the rate of depreciation. Alternatively, the more common story is that rigidities, say, in labor markets cause quantity adjustments in manufacturing industries while time lags in agricultural production cause price adjustments.

For example, fixed wage rates make layoffs a more profitable response to reduced demand for automakers and steel mills than lowering prices since input prices cannot be bid down. Thus, fixed wage rates negotiated by labor unions for the intermediate run can cause short-run market failure by imposing excessive price adjustments on agriculture and excessive quantity adjustments on labor. The adverse effects of this market failure are greater for greater variations in prices or quantities which, in turn, are due to greater variations in macroeconomic policy. When policy makers are not provided with information on how the social costs of these market imperfections are altered by macroeconomic policy, macroeconomic policy externalities are unavoidable. Nevertheless, such information has not been produced in the economics literature.

One might also note that positive consequences could result from extreme macroeconomic instability in this case, as in the case of capital market imperfections, by inducing institutional reform. For example, if overshooting is due to rigidities in input markets for manufacturing, then some of the reduction in market power for labor unions that occurs during periods of instability may reduce the vulnerability of agriculture to overshooting.

Neither of these positive or negative consequences can be fully taken into account in formulating macroeconomic policy on the basis of standard macroeconomic variables such as aggregate inflation and unemployment because the distributional details of aggregation matter. For example, if a change in policy eventually causes a 10 percent reduction in all wages and prices, social welfare considerations are much different than in the short run where agricultural prices might decline by 30 percent while other wages and prices are unchanged. This can be shown formally in a model that aggregates welfare over several markets because the aggregate welfare effects of a change in demand are generally better when both prices and quantities are free to adjust then if only one or the other can adjust.

INTERSECTORAL TRANSMISSION AND DUTCH DISEASE

To this point, the discussion in this paper has focused on macroeconomic externalities that can occur within a single national economy. It now turns to consideration of intercountry macroeconomic externalities. The macroeconomic externalities discussed to this point are externalities that should be considered and can be internalized in national macroeconomic policy formulation. The remaining externalities can be considered in two different spheres of policy formulation. From the standpoint of a national economy, they could be considered in taking maximum advantage of the world economy. On the other hand, from a global standpoint, they should be taken into account in formulating international agreements to maximize social welfare for the world as a whole.

One such consideration arises because of the intersectoral interdependence that occurs with flexible exchange rates. Under flexible exchange rates, a sharp change in economic activity in one sector can cause a change in the exchange rate which in turn alters the effective world price faced by all other traded sectors. This can lead to a world-wide misallocation of resources.

For example, consider the Mexican discovery of oil that occurred in the 1970s. The resulting flow of capital to Mexico induced by the discovery caused the Mexican peso to appreciate rapidly. This then had a negative effect on other tradable sectors because the price of their goods in terms of foreign currency increased (both export demand and import supply declined). This effect fell primarily on agriculture because agriculture had previously been the major export sector. This is the typical story of the "Dutch disease" which has affected several countries since flexible exchange rates have dominated the world economy (see, e.g., Eastwood and Venables, 1982, or Buiter and Purvis, 1983).

The Dutch disease phenomenon can lead to world misallocation of resources because of the conflicting production and consumption signals that occur across international borders. For example, for a time the declining price of wheat in Mexico associated with the oil discovery encouraged a reduction in production while increasing wheat prices just across the border in the United States were encouraging an increase in production. A similar but opposite situation has occurred relative to Canada recently with the declining world markets for potash and uranium. That is, U.S. wheat producers have suffered a much greater weakening of their market than Canadian wheat producers just across the border. Nevertheless, the real resources used by each are essentially the same. Thus, from the standpoint of world production efficiency, one must ask why the two should receive conflicting signals. (Note that identical signals is consistent with the theory of comparative advantage because the factor price equalization theorem implies that individual producers in different countries with identical production functions will face identical prices and, thus, behave identically.)

With better exchange rate management, consequences would still occur in agriculture because resources would tend to be bid away from agricultural production while energy related input prices may fall and consumer incomes may increase. However, these adjustments can be accommodated in line with world markets thus affecting producers abroad in a similar fashion. For example, one way of neutralizing the adverse effects of a booming export sector on nonbooming tradeable sectors is to follow a conscious policy of capital exports. This approach maintains stable trading relationships for other sectors both domestically and abroad because it roughly neutralizes exchange rate effects. In this sense, it is suggestive of the type of international macroeconomic policy coordination that is needed to maintain a stable international economy. However, the necessary timing of capital exports is crucial and few countries have been successful in attaining it (Neary and van Wijnbergen, 1984). The sensitivity of these outcomes to macroeconomic policy and the availability of information for proper timing further illustrates the potential for macroeconomic externality.

PROTECTIONISM

One of the most common domestic policy responses to international instability has been to insulate some domestic prices from movements in world prices or to raise some domestic prices above world levels. For example, the European Community's (EC) variable levies insulate agricultural producers from world price movements. Similarly, U.S. price supports insulate U.S. producers from world prices. At times, the extent of these distortions is not large but with greater international instability the likelihood of occasionally imposing very large distortions is greater. That is, a formal result from welfare economics is that the welfare efficiency loss associated with a price or quantity distortion increases at an increasing rate in moving away from equilibrium; thus, efficiency loss at the average distortion is less than the expected loss with (macroeconomic) instability.

One must also bear in mind that while these sector policies may attain greater stability for decision makers in international markets, this benefit comes at the expense of significant international externalities. That is, producers in one country affect world markets and producers in other countries in ways that are not reflected in their market. For example, U.S. and European producers may continue to produce at high levels during world surpluses causing price reductions abroad even though their domestically supported prices are relatively unaffected.

To the extent that these measures are adopted in response to macroeconomic instability or to the extent that larger distortions tend to occur with macroeconomic instability, these international market failures become macroeconomic externalities. For example, the high valued U.S. dollar resulting from macroeconomic policies has been blamed for the decline in world demand for U.S. grains which in turn, caused the distorting influence of U.S. price supports to increase sharply in the early 1980s. Thus, to the extent that it was not taken into account in formulating macroeconomic policy, the increased distorting influence can be viewed as a macroeconomic externality.

A more common form of protection in developing countries is to impose trade duties on imports of industrial goods which compete with domestic manufacturers. As a result, the prices of protected tradable industrial goods increase relative to the prices of traded agricultural goods and nontradeable goods thus causing an improved overall trade balance. With flexible exchange rates, this tends to turn the exchange rate against agricultural exports (Little, Scitovsky, and Scott, 1970; Krueger, 1978; Schultz, 1978). The effect of protection on agriculture has been shown to be important empirically by Garcia (1981) in Colombia and by Cavallo and Mundlak (1982) in Argentina. Garcia found, for example, after equilibrium adjustments of the exchange rate in response to protection, that a uniform tariff of 30 percent on all imports imposed a tax equivalent of 27 percent on all exports.

This type of protection is essentially a macroeconomic policy aimed at fostering growth of the domestic (industrial) economy. Thus, the corresponding adverse effects on export sectors, to the extent they are not taken into account in policy formulation, are macroeconomic externalities. Most evidence suggests that these effects have not been taken into account in many countries. That is, even though protectionist policies were followed to increase economic growth, many empirical studies have found that import substitution slows economic growth by constraining international exchange (see Bhagwati and Krueger, 1973; Krueger, 1983). In contrast, many of the developing countries that have achieved rapid growth have followed strategies of export promotion.

ENDOGENOUS FOREIGN MACROECONOMIC POLICY RESPONSE

One of the most significant changes that occurred with the collapse of the Bretton Woods Agreement was that countries became free to try to export domestic problems related to inflation, unemployment, and external accounts through monetary and fiscal policy. However, some countries are too large relative to world markets for the related effects on other countries to go unnoticed. Policies undertaken by the major economies affect the rest of the world through the integrated capital market and associated exchange rate adjustments.

For example, a budget deficit arising from a tax cut causes a higher relative price of nontradeable goods and higher interest rates which, in turn, lowers foreign wealth and consumption and reduces the foreign relative price of nontradeable goods. Frenkel and Razin (1986) argue that these theoretical results explain the effects of U.S. fiscal policies of the early 1980s on Japan, Germany, and the United Kingdom and the associated exchange rates. However, while the declining value of European currencies caused the European Community's trade balance to improve rapidly, it also caused a doubling of inflation. This induced a foreign policy response.

To limit further inflation, the German money supply, for example, was actually reduced between 1980 and 1981 which caused unemployment to double and real growth to decline sharply. The associated high interest rates and slack economy also caused a tendency to increase the budget deficit to which the German government responded by reducing spending. Of course, these revisions in European macroeconomic policy had further feedback effects on the United States. These adverse consequences and the necessary policy responses explain why European policy officials have been unanimous in calling upon the United States to reduce its budget deficit and cooperate in intervention to lower the dollar's value (Feldstein, 1986). However, they also exemplify another potential type of macroeconomic externality.

When changes in macroeconomic policy induce endogenous changes in macroeconomic policies of other countries, they must be taken into account. If they are not, or if incorrect judgments are made about those responses, then policy makers will not properly account for the true marginal social impact of their decisions. Macroeconomic externalities will thus occur.

One such possibility that must be considered is whether the increased macroeconomic instability among developed countries has induced increased protectionist policies among developing countries. Protection causes world market failure and to the extent that it is induced by macroeconomic policies formulated abroad without regard to these effects, the adverse consequences of this market failure are a macroeconomic externality.

Another possibility that deserves consideration relates to world capital market failure. That is, the macroeconomic policies of developed countries may affect the ability of developing countries to finance government spending by borrowing. For example, with the oil boom many oil-producing countries attempted to invest petrodollars abroad which made credit readily available and induced heavy borrowing among developing countries. Then as money supply growth was reduced sharply in the early 1980s (mainly in the United States), interest rates increased rapidly leaving several countries unable to service their debt and/or unable to obtain further credit. This in turn affected the ability of these countries to buy U.S. agricultural exports and imposed many other costs of adjustment both domestically and abroad.

The extent to which macroeconomic policy reactions and adjustments abroad are taken into account is unclear but with little doubt they are not fully considered. Barro and Gordon (1983), for example, have found an inflationary bias in policy formulation. Furthermore, in many cases, responses abroad become clear only after long periods of debate once the associated circumstances are realized. Also, many cases reveal that expectations have not been realized. For example, President Nixon clearly had long term hopes for the Smithsonian Agreement of December 1971 when he hailed it as "the most significant monetary agreement in the history of the world." But when the United States continued with its overly expansionary monetary policy which had caused the collapse of the Bretton Woods Agreement, the associated exporting of inflation to Germany and Japan led to its collapse as well after only 14 months. As a result, monetary policies in major countries became quite erratic and often exacerbated by one another's policy adjustments.

Such examples raise the prisoner's dilemma issue. Perhaps a loss of economic welfare occurs for all countries as a result of individual countries trying to take advantage of the rest of the world. McKibbin and Sachs (1986) demonstrate such a possibility where each country tries to export a world inflationary shock by appreciating its currency. The result is excessive monetary contraction and fiscal expansion. These circumstances roughly approximate the international economy following the oil price shocks of the 1970s and thus suggest the presence of international macroeconomic externalities. Agriculture is a particularly critical loser from this kind of world macroeconomic instability because of the overshooting that tends to occur in agricultural markets.

ENDOGENOUS SECTOR POLICY RESPONSE

Policy makers must also bear in mind the two-way interaction that occurs between sector policies and macroeconomic policies. For example, many governments subsidize agriculture more, or tax it less, when agricultural markets are weak. This is clearly the case with the European variable trade levies. It also tends to be the case with U.S. price supports, but they are often revised in response to perceived long term conditions. In many developing countries, such as Colombia with coffee and Thailand with rice, trade duties are revised to compensate somewhat for changing world market conditions (Weisner, 1978). Of course, these agricultural subsidies and taxes have a direct effect on government deficits. As a result, lower prices in world agricultural markets can lead to larger fiscal deficits due to endogenous sector policy adjustments.

These effects give rise to the potential for macroeconomic externalities on several levels. First, policy makers must consider the effects of macroeconomic policy on the behavior of existing domestic sector policies. This may be a relatively reasonable task for policies such as the European variable trade levies where automatic adjustments are prescribed.

Second, policy makers must consider changes in sector policies that may be required to deal with resulting circumstances. Such changes have been frequent in U.S. agricultural policies because of their traditional inflexible form (Just and Rausser, 1984). These changes, which may become politically necessary as the result of interest-group lobbying, are more difficult to forecast. For example, the U.S. Soybean embargo in June 1973 was imposed in response to a booming export market that threatened domestic shortage. This occurred in large part because macroeconomic policies led to a sharp decline in the value of the dollar and was apparently an unexpected effect. Thus, both the effects of the shortage and the efficiency losses associated with the embargo were macroeconomic externalities.

A third effect that policy makers must consider is the adjustment of foreign sector-specific policies induced indirectly by macroeconomic policy. For example, the high value of the U.S. dollar in the early 1980s allowed Argentina to adopt expansionary wheat and corn policies, Brazil to adopt expansionary soybean policies, and Thailand to adopt expansionary rice policies, all of which displaced U.S. exports. To the extent that these foreign sector policy adjustments were not taken into account in formulating U.S. macroeconomic policies which caused the dollar to appreciate, macroeconomic externalities occurred that were primarily harmful to U.S. farmers.

POLICY STICKINESS

In the case of each of the types of policies discussed here, policy stickiness may also be an important issue. For example, many countries use external shocks as a reason for instituting various kinds of policies. However, once instituted, policy controls have a tendency to become permanent and remain after the shocks have disappeared. Vested interest groups tend to organize and, once developed, can lobby effectively against discontinuation. As a result, policy instruments often fall into misuse or begin to be used for other purposes for which they are not well suited. To the extent that these effects occur as an unanticipated effect of macroeconomic policies, they cause macroeconomic externalities.

In the policy environment of the United States, these effects can be strong and long lasting. For example, it has often been argued that the Great Depression lies behind the structure of U.S. agricultural policy that has existed since then. Thus, the distorting effects of the last 50 years of agricultural policy may be largely an externality associated with the macroeconomic policies that caused the Depression.

INFORMATION AND TRANSACTION COSTS

A country that fixes the exchange rate of its currency vis-a-vis another currency is generally regarded as granting a positive externality to the users of the other currency (see, e.g., Vaubel, 1984). Likewise, exchange rate adjustment and instability causes negative externalities. In addition to the sources of externalities discussed above (e.g., risk), these externalities arise from higher transactions costs associated with currency conversion and increased information requirements.

With exchange rate instability, firms dealing in international markets or producing for export markets must be concerned not only with the potential for shocks to their commodity markets between the times of decisions and actual transactions, but also with shocks to the associated exchange rates. This may necessitate hedging in futures markets and monitoring some of the wide range of phenomena that may affect exchange rates (e.g., foreign macroeconomic policy and the economic conditions of other major industries that affect floating exchange rates through trade and capital movements). These activities dictate the development of expertise by individual firms in many more areas than with fixed exchange rates. All of these activities represent increased transactions and information costs associated with exchange rate instability.

To the extent that these costs are not taken into account by macroeconomic policy makers in setting policies that affect exchange rate stability, they represent macroeconomic externalities. If macroeconomic policy is formulated on the basis of macroeconomic variables, however, these costs tend to be miscounted because they contribute positively to national income associated with the value of information under exchange rate instability. With exchange rate stability, on the other hand, the productivity of resources used to produce this information could be captured in their next best uses without loss to the industry because the related information costs would be unnecessary. These considerations are particularly relevant for agriculture because the demand for farm products tends to be exchange rate elastic in the short run and because production lags are large. Thus, the premium for such information is greater.

CONCLUSIONS: PROSPECTS FOR REDUCING MACROECONOMIC EXTERNALITIES

This paper enumerates a number of market failures and externalities that are caused or exacerbated by macroeconomic policy instability particularly when it is formulated strictly on the basis of macroeconomic indicators. The general implication of the paper is that macroeconomic policy formulation should take these considerations, many of which have adverse consequences for agriculture, into account. In some cases, the arguments suggest greater consideration of induced policy responses both domestically and abroad at both the sector and macro levels. In other cases, the arguments suggest greater consideration of the market failures that make macroeconomic indicators of the economy inadequate guides for policy formulation. Some may argue that these effects, whether they are called macroeconomic externalities or not, are already considered. On the contrary, the lack of sound empirical economic research on many of these issues implies that economists know little about their magnitude. Many have received theoretical or conceptual attention in the literature only recently. Thus, policy makers are not being provided with the necessary information for adequate consideration.

When better information is produced about how foreign macroeconomic policies are affected by domestic macroeconomic policy revisions, the tendency toward prisoner's dilemma possibilities may be substantially reduced. For example, when knowledge about foreign reactions makes the adverse consequences of attempting to export domestic problems apparent, the incentive for international cooperation may be increased.

One recent event that suggests that the costs of uncoordinated international macroeconomic policy may have become so great as to induce beneficial reforms is given by the September 1985 G-5 accord (among the United States, United Kingdom, Germany, France, and Japan). This agreement committed the major economies of the world to some degree of monetary policy coordination for the purpose of managing exchange rates (mainly the value of the dollar) that have apparently been successful. However, the nature of these commitments is somewhat loose especially with respect to fiscal policy and has apparently led to some exporting of fiscal deficits from the United States to Germany and Japan. Sachs (1986) argues that this problem is responsible for a lack of enthusiasm by Germany and Japan that may lead to collapse of the agreement much as in the case of the Smithsonian Agreement in 1971-73. He argues that exchange rate management per se will probably be ineffectual beyond the short run unless new rules of the game are determined to coordinate national policies more completely.

The reason for a sustainable framework of exchange rate management centers around the importance of international money. Mundell (1985) and Kindleberger (1981) argue that what stability has existed in the international economy over the last 15 years is due to the dominance of the dollar in world trade (rather than the efficiency of a multicurrency world). The United States has been willing to bear the costs of providing this international public good, perhaps because of the positive externalities enjoyed by Americans when a larger share of the world is on a dollar standard. However, the U.S. economy is shrinking as a share of world income, has recently become a net debtor nation, and is becoming increasingly vulnerable to external shocks. These increasing costs may cause the United States to shrink from its role in providing some form of international money perhaps because of a loss in confidence in the dollar. If so, continued provision of some form of international money may depend on whether other countries become willing to cooperate in its provision (which will apparently require coordination of national policies through some new institutional arrangement).

Many of the macroeconomic externalities discussed in this paper could be reduced or eliminated if these possibilities were realized. For example, this could eliminate the need for protection from instability in export demand due to short run exchange rate adjustments in response to monetary and fiscal policies that are incompatible in the long run either intranationally or internationally. For this reason, some have argued that reform of international monetary arrangements should be at the top of the agricultural policy agenda (Schuh and Orden, 1985). Alternatively, one must question the likelihood of such arrangements and the likelihood that agricultural concerns can be successful in lobbying for such a broad scale change.

Another approach for dealing with several of the macroeconomic externalities discussed in this paper is through flexible policies with automatic adjustments imposed through legislation. Once the objectives of domestic agricultural policy are formalized, it is immediately evident that revisions in agricultural policy are required to meet the objectives as major revisions occur in macroeconomic policy, both at home and abroad (Just, 1984a). The potential gains from tying agricultural policy instruments to macroeconomic variables appear sizeable for several reasons.

First, the tendency for temporary policies (adopted in response to external shocks) to become permanent is reduced by legislatively tying policy instruments to the shocks which they are intended to address. Likewise, this kind of policy formulation can help policy instruments respond more quickly to developing crises (Just and Rausser, 1984). Given that macroeconomic instability is to continue, these measures can reduce the macroeconomic externalities associated with agricultural policy stickiness (Just, 1984b).

A second consideration has to do with expectations. With inflexible policies that must be frequently revised in periods of instability, producers may not expect that a policy will continue so they may be reluctant to respond to its inducements. Alternatively, they may incorrectly expect certain policies to continue and thus find themselves poorly suited to adjust to their removal. In either case, producers may incur costs of adjustment that could be avoided with more gradual policy changes that can be anticipated on the basis of observable economic phenomena and legislatively specified rules (Just, 1984b).

Another potential benefit of tying agricultural policy instruments to macroeconomic variables is that it tends to make macroeconomic policy makers and policy analysts as well as private decision makers recognize explicitly the interactions that exist. This directly reduces some of the potential for macroeconomic externalities whereby induced changes in sector policy are not fully considered in macroeconomic policy formulation. It also reduces the agricultural policy uncertainty faced by farmers which can be another externality of macroeconomic instability (Just and Rausser, 1984).

Some of the remaining macroeconomic externalities considered in this paper arise because of existing market failures (capital market imperfections, risk aversion, and possibly overshooting). In these cases, one could blame these social costs on the traditional forms of market failure and argue that the fact that they increase with macroeconomic instability is further justification why they should be removed through (termination of) other policies. However, not all of the market failures discussed here can be eliminated efficiently through other policies (without excessive social cost). For example, the problem of moral hazard makes some kinds of social cost associated with risk aversion appropriate to incur.

More properly, macroeconomic policy formulation should take into account how the distortions associated with these failures are affected and whether correcting policies should be adopted in conjunction with macroeconomic policy revision. These considerations should take into account the extent to which such failures are controllable and the extent to which the effects of such controls are understood. Alternatively, macroeconomic policy formulation must take a second best approach whereby these market imperfections are taken as given. Given these imperfections, the classical efficiency arguments for perfectly flexible exchange rates made, for example, by Friedman (1953) may be inapplicable. In a second best context, macroeconomic policy formulation becomes an exceedingly complex problem.

When macroeconomic policy is formulated on the basis of macroeconomic variables in an economy free of distortions, maximization of economic growth and real national income can be viewed as equivalent to or an approximation of maximization of the welfare economist's concept of social welfare under certain conditions (Harberger, 1971; Sen, 1979). However, maximization of real national income can fail to optimize social welfare for a variety of reasons (Samuelson, 1950). In particular, when a policy is changed in an economy with existing distortions, the change in national income must be augmented by the change in marginal social benefit minus marginal social cost multiplied by market quantity in each distorted market (Just, Hueth, and Schmitz, 1982, Appendix D).

Empirical research evaluating the effects of macroeconomic policy on the welfare effects of existing distortions is so sadly lacking that it is difficult to imagine these effects have been properly considered. Research is needed on the effects of both monetary and fiscal policies and of the resulting exchange, inflation, and interest rates on distorted markets and their welfare implications. Several obvious candidates in agriculture include agricultural credit markets (because of capital market imperfections), contingency markets for transferring risk (because of risk aversion and the failure of crop insurance due to moral hazard), regulated input and output markets (because of the prevalent role of pesticide regulations, output price support and subsidy programs, import regulations, and occasional embargoes), and input and output markets that may be subject to market power (such as farm machinery and major grain and meat markets). Without better understanding of how macroeconomic policy affects the social costs of these problems, policy makers cannot be expected to formulate macroeconomic policy that fully accounts for agricultural interests. The blame for not providing this information rests squarely at the feet of the agricultural economics profession.

REFERENCES

Arrow, K. J., and R. C. Lind. "Uncertainty and the Evaluation of Public Investment Decisions," American Economic Review. 60(1970): 364-378.

Barro, R., and D. Gordon. "Rules, Discretion, and Reputation in a Model of Monetary Policy," Journal of Monetary Economics. 12(1983): 101-121.

Bhagwati, J. N., and A. O. Krueger. "Exchange Control, Liberalization, and Economic Development," American Economic Review. 63(1973): 419-27.

Brainard, W. C., and R. N. Cooper. "Uncertainty and Diversification in International Trade," Food Research Institute Studies. 8(1968): 257-285.

Bryant, R. C. Money and Monetary Policy in Independent Nations. Washington, D.C.: The Brookings Institution, 1980.

Buiter, W. H., and D. D. Purvis. "Oil, Disinflation and Export Competitiveness," Economic Interdependence and Flexible Exchange Rates. J. Bhandari and B. Putnam, (eds.). Cambridge, MA: MIT Press, 1983.

Cavallo, D. F. "Exchange Rate Overvaluation and Agriculture," background paper for WDR 86, World Bank, Washington, D.C., September, 1985.

______, and Y. Mundlak. "Agriculture and Economic Growth in an Open Economy: The Case of Argentina," Research Report 36, International Food Policy Research Institute, Washington, D.C. December, 1982.

Chetty, K. V. "On Measuring the Nearness of Near-moneys," American Economic Review. 59(1969): 270-281.

Chambers, R. G. and R. E. Just. "Effects of Exchange Rate Changes in U.S. Agriculture: A Dynamic Analysis," American Journal of Agricultural Economics. 63(1981): 32-46.

Dornbusch, R. "Expectations and Exchange Rate Dynamics," Journal of Political Economy. 84(1976): 1161-76.

Eastwood, R. K., and A. J. Venables. "The Macroeconomic Implications of a Resource Discovery in an Open Economy," Economic Journal. 92(1982): 285-299.

Feldstein, M. "U.S. Budget Deficits and the European Economies: Resolving the Political Economy Puzzle," American Economic Review. 76(1986): 342-346.

Fischer, S. "Money and the Production Function," Economic Enquiry. 12(1974): 517-533.

Frenkel, J. A., and A. Razin. "The International Transmission and Effects of Fiscal Policies," American Economic Review. 76(1986): 330-335.

Friedman, M. "The Case for Flexible Exchange Rates," Essays in Positive Economics, Chicago: University of Chicago Press, 1953.

_______. A Program for Monetary Stability. New York, 1959a.

_______. "The Demand for Money: Some Theoretical And Empirical Results," Journal of Political Economy, 67(1959b): 327-351.

_______. The Optimum Quantity of Money and Other Essays. Chicago, 1969.

Fry, M. J., and D. M. Lilien. "Monetary Policy Responses to Exogenous Shocks," American Economic Review. 76(1986): 79-83.

Garcia, J. The Effects of Exchange Rates and Commercial Policy on Agricultural Incentives in Colombia: 1953-1978. Research Report 24, International Food Policy Research Institute, Washington, D.C., June, 1981.

Hall, R. E. The Role of Government in Stabilizing Prices and Regulating Money. Stanford University, working manuscript, 1981.

Harberger, A. C. "Three Basic Postulates for Applied Welfare Economics: An Interpretive Essay," Journal of Economic Literature. 9(1971): 785-797.

Just, R. E. "Automatic Adjustment Rules for Indexing Support Prices," Department of Agricultural and Resource Economics, University of California, Berkeley, 1984a.

_______. Automatic Adjustment Rules for Agricultural Policy Controls, American Enterprise Institute for Public Policy Research, Studies in Economic Policy, Washington, D.C., November, 1984b.

_______. "Risk Aversion Under Profit Maximization," American Journal of Agricultural Economics. 57(1975): 347-352.

_______, and G. C. Rausser. "Uncertain Economic Environments and Conditional Policies," Alternative Agricultural and Food Policies and the 1985 Farm Bill. G. C. Rausser and K. R. Farrell (eds.), Giannini Foundation of Agricultural Economics, University of California, Berkeley, 1984, pp. 101-132.

Just R. E., D. L. Hueth, and A. Schmitz. Applied Welfare Economics and Public Policy. New York: Prentice-Hall, 1982.

Khan, M. S. "Developing Country Exchange Rate Policy Responses to Exogenous Shocks," American Economic Review. 76(1986): 84-87.

Kindleberger, C. International Money. London: Allen & Urwin, 1981.

_______. "International Public Goods without International Government," American Economic Review. 76(1986): 1-13.

Klein, B. "The Competitive Supply of Money," Journal of Money, Credit, and Banking. 6(1974): 423-453.

Kolm, S. C. "External Liquidity--A Study in Monetary Welfare Economics," Mathematical Methods in Investment and Finance. G. P. Szego and K. Shell (eds.), Amsterdam/London/New York, 1972: 190-206.

Kormendi, R. C., and P. G. Mequire. "Cross-Regime Evidence of Macroeconomic Rationality," Journal of Political Economy. 92(1984): 875-908.

_______. "Macroeconomic Determinants of Growth: Cross-County Evidence," Journal of Monetary Economics. 16(1985): 141-163.

Krasner, S. D. International Regimes. Ithaca: Cornell University Press, 1983.

Krueger, A. O. "Protectionism, Exchange Rate Distortion, and Agricultural Trade Patterns," American Journal of Agricultural Economics. 65(1983): 864-71.

_______. Foreign Trade Regimes and Economic Development: Liberalization Attempts and Consequence. Cambridge, MA: Ballinger Publishing Company, 1978.

Little, I. M. D., T. Scitovsky, and M. Scott. Industry and Trade in Some Developing Countries. New York: Oxford University Press, 1970.

Longmire, J. and A. Morey, "Strong Dollar Dampens Demand for U.S. Farm Exports," FAER No. 193, Economic Research Service, U. S. Department of Agriculture. Washington, D.C. 1983.

McKibbin, W., and J. Sachs. "Coordination of Monetary and Fiscal Policies in the OECD," International Aspects of Fiscal Policies. J. Frenkel (ed.), National Bureau of Economic Research, forthcoming, 1986.

Mundell, R. "Proposals for the Future International Economic System," presented at the Congressional Summit on Exchange Rates and the Dollar. Washington, D.C., November, 1985.

Mussa, M. "A Model of Exchange Rate Dynamics," Journal of Political Economy. 90(1982): 74-104.

Neary, J. P., and S. van Wijnbergen. "Can an Oil Discovery Lead to a Recession? A Comment on Eastwood and Venables," Economic Journal. 94(1984): 390-395.

Niehans, J. The Theory of Money. Baltimore/London, 1978.

Phelps, E. S. and J. B. Taylor. "Stabilizing Powers of Monetary Policy Under Rational Expectations," Journal of Political Economy. 85(1977): 164-190.

Rausser, G. C. Macroeconomics of U. S. Agricultural Policy. Studies in Economic Policy, American Enterprise Institute for Public Policy, Washington, D. C., 1986.

Sachs, J. "The Uneasy Case for Greater Exchange Rate Coordination," American Economic Review. 76(1986): 336-341.

Samuelson, P. A. "Evaluation of Real National Income," Oxford Economic Papers. (New Series), 2(1950): 1-29.

Schuh, G. E. "Strategic Issues in International Agriculture," Working Draft, Agriculture and Rural Development Department, World Bank, Washington, D.C., 1985.

________, and D. Orden. "The Macroeconomics of Agriculture and Rural America," paper presented at the Conference on Agriculture and Rural Areas Approaching the 21st Century: Challenges for Agricultural Economics, sponsored by the American Agricultural Economics Association. Ames, Iowa, August 7-9, 1985.

_______. "The Exchange Rate and U. S. Agriculture," American Journal of Agricultural Economics. 56(1974): 1-13.

Schultz, T. W. Distortions of Agricultural Incentives. Bloomington, Ind: Indiana University Press, 1978.

Sen, A. "The Welfare Basis of Real Income Comparisons: A Survey," Journal of Economic Literature. 17(1979): 1-45.

Stamoulis, K. G., J. A. Chalfant, and G. C. Rausser. "Monetary Policies and the Overshooting of Flexible Prices: Implications for Agricultural Policy," Department of Agricultural and Resource Economics, University of California, Berkeley, 1985.

Valdes, A. "Exchange Rates and Trade Policy: Help or Hindrance to Agricultural Growth?" paper presented at the XIX International Conference of Agricultural Economists, Malaga, Spain, August 6-September 4, 1986.

Vaubel, R. "The Government's Money Monopoly: Externalities or Natural Monopoly?" Kyklos. 37(1984): 27-58.

Weisner, E. Politica Monetaria y Cambiaria en Colombia. Bogota: Association Bancaria de Colombia, 1978.

Willett, T. D. "Macroeconomic Instability and Exchange Rate Volatility." Claremont, Calif: Claremont Graduate School, 1981.

7
U.S. Macroeconomic Policy and Agriculture

Robert L. Thompson

Macroeconomics and agriculture are neither new nor particularly strange bedfellows. In the late 19th century, William Jennings Bryan assailed the gold standard, arguing for free coinage of silver to unleash farmers from the depressed conditions in which they found themselves. He argued that U.S. farmers were being crucified on a "cross of gold". While Bryan's candidacy for the presidency was unsuccessful, his argument eventually led to reform of the monetary system.

From Colonial times through the late 19th and early 20th century, agriculture was the principal earner of export revenue for the United States. Exports of indigo, tobacco, cotton, and naval stores all were vital to the U.S. economy. This role declined as we moved into the twentieth century, and the 40 years from 1933 through 1973 saw exports become a relatively unimportant outlet for farm production. Commodity programs became the dominant form of public policy affecting farmers' well being. This began to turn around again, however, as we entered the decade of the 1970s.

The fraction of total farm output which is exported more than doubled over the 1970s, resulting in the exchange rate becoming a more and more important price affecting farmers' well-being. This effect was reinforced by the increasing volatility of exchange rates in the 1970s, when it appeared that as exchange rates varied, so varied agricultural exports.

Robert L. Thompson, Dean of Agriculture, Purdue University. At the time of these remarks Assistant Secretary of Agriculture for Economics. U. S. Department of Agriculture.

Agriculture became one of the more capital intensive sectors of the U.S. economy, and the total farm debt load more than quadrupled in the 1970s to the point that the interest rate became the single most important price affecting net farm income.

Also, following a period when food prices generally had been a stabilizing factor in the general price level, in 1973, food prices rose at more than twice the general rate of inflation in the rest of the economy. In the 1980s, the opposite has occurred. As inflation has slowed, food price increases have slowed even more.

Thus, there are three primary linkages between macroeconomics and agriculture which we'll look at here: cyclical linkages; interest rates; and exchange rates.

CYCLICAL LINKAGES BETWEEN MACROECONOMIC POLICY AND AGRICULTURE

Traditionally, we tend to view agriculture as being weakly linked to the level of macroeconomic activity because, in a mature economy, income elasticities of demand for farm products are low. Thus, any change in national income induced by macroeconomic policy tends to have a small effect on the farm sector. The income-consumption linkage is stronger for meats and livestock products than for most crops, but the elasticities are still not as large as those affecting many other products elsewhere in the economy.

Nevertheless, despite the fact that agricultural product income elasticities are low, short run price elasticities of supply are even lower. As a result, macroeconomic policy-induced shifts in demand for farm products can have significant price effects. Moreover, income elasticities of demand for agricultural products are much higher in the lower income countries we sell to. This means that cyclical effects can get translated even more strongly into demand shifts through foreign markets.

Probably even more important, however, is the fact that, except when sitting on support levels, agricultural prices tend to be flexible in a fairly short period, whereas prices of many goods produced by the manufacturing sector of the economy are stickier in the short run due to longer contract periods. There's little short run flexibility in these other prices in response to monetary shocks. On the other hand, prices of agricultural products, together with other flexibly-priced goods, tend to adjust more than proportionately, or overshoot in the short run. That is, monetary shocks are not neutral at least in the short run. The events of the 1970s can easily be viewed as overshooting on the upside in response to a rapid expansion in global and U.S. liquidity, starting with the guns-and-butter policy of the late 1960s when we financed the Vietnam War without raising taxes but, rather, by printing money. This overshooting in commodity prices led to a land price boom--especially in the most export dependent regions of U.S. agriculture--and set off massive capital investment in agriculture.

The converse appears to have been the case in the early 1980s when, as a result of the stringent monetary policy implemented to get double-digit inflation under control, agricultural prices weakened proportionately more than other prices in the economy until they hit the loan rates written into the 1981 farm bill. This downturn caused the speculative bubble in the land market to burst, and negative net investment in agriculture--as characterized by 20% rate of use of capacity in farm machinery manufacturing. It appears that agriculture has become one of the sectors which is subjected to larger adjustments than others in the economy.

As a result, there are those who have suggested, at first facetiously but now more seriously, that perhaps the greatest justification for farm policy as we now know it is to insulate agriculture from part of the shocks it endures from macroeconomic policy--or at least to offset some of these shocks. In fact, the 1981 farm bill may inadvertently have written into law automatic stabilizers that protected U.S. farmers from much worse problems that would have come from the overshooting that would have occurred. That that farm bill cost $52 billion more than projected illustrates that such a policy can be very costly.

Now, I want to go back and discuss the interest rate and the exchange rate, the other two principal macroeconomic linkages.

INTEREST RATES

In recent decades, agriculture has become one of the more capital-intensive sectors of the U.S. economy. By the end of the 1970s, agriculture used twice as much machinery and equipment per person employed as the manufacturing sector. The capital-output ratio in agriculture was three times that of the U.S. economy as a whole by decade's end. Moreover, agriculture also has become increasingly dependent on debt financing. During the 1970s the total debt load of American farmers increased from $65 billion to more than $200 billion.

As a result, the interest rate really became the single most important price affecting farmers' bottom line. In 1984, American farmers paid $21 billion in interest on borrowed money--six times the $3.4 billion spent in 1970. Interest rates have another less visible role. The interest rate is the opportunity cost of capital tied up in land, inventories, livestock, and the like. Therefore, with the increasing integration of capital markets and the increasing real interest rate charged to farmers in recent years, the interest rate has become an even more important determinant of the value of these variables and the change in their investment.

In previous decades, when real interest rates often were negative and generally less than 4 percent, farmers could perhaps afford to pay little attention to the opportunity cost of capital. Moreover, regulations in the banking industry tended to isolate rural credit markets from the national money market, further adding to rural credit market stability. However, when real interest rates in the early 1980s rose to unprecedented levels of 5-6--and occasionally as high as 10-12--percent, interest rates suddenly became a much more important determinant of how large a herd of cattle a farmer kept, how large a stock of grain he carried over, and--probably even more important--how much he was willing to pay for land.

This increase in real interest rates was one of the major factors causing the downward spiral in land prices in the 1980s. Increased variability in interest rates also has added another source of riskiness to farm income and investment. This likely has reduced the optimum debt load a farmer can carry. This could lead in the future to increasing reliance by farmers on equity financing as opposed to debt financing. Banking deregulation creates stronger links between rural interest rates and interest rates at money-center banks. Removal of interest-rate ceilings and relaxing of rules on interstate banking and portfolio regulation have broken the isolation of rural credit markets from the national market. Although this means rural areas feel monetary shocks more quickly, deregulation may also increase diversification and access to funds that could reduce the cost of capital in rural areas.

EXCHANGE RATES

The exchange rate became an important variable affecting the well being of agriculture in the 1970s, because of both the magnitude of exchange rate movements and U.S. agriculture's growing dependence on exports. This probably would have gone unnoticed had we continued to live under a fixed exchange rate regime through the 1970s. However, this was not the case.

In 1971 and 1973, the U.S. dollar experienced two substantial devaluations. Thereafter, the dollar was cut free to float and find its market-clearing level. Government intervention blunted the swings in exchange rates somewhat through the mid-1970s but by the end of the decade, the U.S. dollar was virtually in free-float.

As a result, macroeconomic policy shocks had a new channel of transmission to the traded-goods sectors of the American economy such as agriculture. Agriculture now is subjected to a large new source of uncertainty.

In addition to moving to floating exchange rates, the 1970s saw a second major change in international financial markets. This was the increasing facility with which capital flowed among countries. Barriers to international capital flows which had existed for decades (maybe forever) were removed, and technological change in international telecommunications created an environment in which billions of dollars literally can be transmitted across national borders at the touch of a telex key.

This is an important part of our story because international capital flows have become increasingly sensitive to small changes in expected returns in different countries, and macroeconomic policy changes are the principal determinant of these changes in expected returns. The growth in world capital markets of the 1970s and 1980s has had a profound impact on exchange rate determination. Financial markets have grown to overshadow goods markets due to both their size and the speed with which financial transactions are made.

To put this in perspective, the dollar value of the goods and services moving into and out of the United States in 1985 exceeded $550 billion. The capital flowing into and out of the United States, however, is estimated to have been as much as $35 trillion. This same size differential is at work in world trade and financial flows. The value of goods and services traded globally in 1984 was approximately $4 trillion, while world financial flows are estimated as high as $200 trillion.

The speed of adjustment also works to increase the importance of financial flows. Bank deposits, government bonds, and stock certificates all change hands far more rapidly than wheat, corn, iron ore, and automobiles. These size and speed considerations suggest that in many cases it could well be financial flows that drive short run exchange rate movements and thereby lead to adjustments in the goods and services market.

This illustrates the extent to which capital flows now dominate trade flows in determining exchange rates. If macroeconomic conditions mandate a large net capital inflow into the United States, under floating exchange rates the market clearing conditions dictate that there be an equal and offsetting current account deficit.

The exchange rate simply has to appreciate by a sufficient margin to ensure that the requisite current account deficit is generated. Therefore, we find the exchange rate being one of the principal sources of transmission of monetary shocks and thereby, one of the principal sources of instability affecting agriculture in the last decade.

If we lived in a world of instant and costless adjustment, this might not be a source of concern. But in the real world, agriculture is a highly capital intensive industry, and the opportunity cost of sunk investments is extremely low. Therefore, these unanticipated and unpredictable exchange rate movements can either generate unanticipated windfall gains or unanticipated low or even negative returns to farm investment. Investments that appeared profitable ex ante on the basis of one set of exchange rate expectations may prove to be disastrous investments ex post if the exchange rate moves significantly against that sector.

This is particularly true for the most fixed of all the resources used in agriculture--land. The land value movements of the 1970s and 1980s reflect to a large extent this phenomenon of ex ante and ex post swings in the extreme. For example, the $2,000-an-acre value of farmland and buildings in Iowa in 1981 had plunged to $840 by early 1986.

The Baker Initiative of September 1985 and the ensuing discussions which resulted in the May 1986 Tokyo Economic Summit Communique acknowledged the adverse effects of the artificially high dollar of the early 1980s and of the volatility of the dollar exchange rate in the last 15 years or so. These communiques called for greater international monetary coordination with the objective of providing somewhat greater stability to exchange rates, and, in turn, to the traded goods sectors of our economy. Deficit reduction measures, such as Gramm-Rudman-Hollings, also should bring about lower exchange rates by reducing the competition between the public and private sectors for the limited supply of domestic savings. This in turn will reduce the need for as large a net capital inflow and thereby the market pressure keeping the trade deficit high to offset that capital inflow. Now that the United States is the world's largest debtor nation, the exchange rate eventually will have to fall to the point that the United States consistently runs a balance of trade surplus in order to make the interest payments on our international debt and to pay down that debt over time.

FISCAL POLICY

Up to this point, I have emphasized the aggregate effects of monetary and fiscal policy on agriculture. Fiscal policy's impact has been mainly through the large federal budget deficits, but we should not ignore the microeconomic effects of fiscal policy on agriculture or the effects of the tax code on investment, supply, and net returns to farmers.

In recent years, commodity programs have transferred something on the order of $15 billion a year to American farmers. Less well known is the fact that for tax purposes, the total losses of American farmers exceed total profits by more than $10 billion. In other words, agriculture has become a large net tax shelter. Certain parts of the farm sector, in particular custom feeding of cattle, dairy, orchards, and vineyards, have been the recipients of significant amounts of nonfarm tax shelter investments. These investments work to expand supply and weaken sensitivity to price changes as investors maximize the tax shelter value of their investments. The resulting larger supply has depressed market prices for these commodities (or contributed to larger government stocks when market prices are sitting on loan rates). This price reduction has depressed returns to family farmers in the business of farming to earn their living, not in the business of farming the tax code.

The Tax Reform Act of 1986 will change this significantly. Most importantly, reducing the maximum marginal tax rate to 28 percent will substantially reduce the incentive for tax shelters in general. In addition, the deletion of investment tax credit, changes in depreciation periods, and changes in capital gains treatment of farm income all reduce the attractiveness of the farm sector in particular for tax sheltering.

The Act explicitly precludes passive nonfarm investors using losses on farm investments to offset nonfarm earnings. These measures all should contribute to improving the supply-demand balance in agriculture by reducing the overproduction now generated by the tax code. Family farmers also will benefit, of course, from the significant reduction in the highest marginal tax rates, from increasing the personal exemption and standard deduction, and from such measures as permitting self-employed people to deduct half of their health insurance premiums.

But the important point here is that fiscal policy can have a major effect on agriculture not only through its aggregate impact on interest rates and exchange rates, but also much more directly through its effect on capital investment and, in turn, on the supply of farm products and farmers' net cash incomes.

MACROECONOMICS AND THE AGRICULTURAL POLICYMAKER

The growing importance of macroeconomic policy relative to farm policy in determining farmers' well-being has changed the Washington environment. Farm groups, accustomed to getting action from Congress and Secretaries of Agriculture to support a price or restrict supply or buy up surplus production to solve farm problems, are tremendously frustrated.

They perceive that interest rates and exchange rates have a dominant influence on their well-being, but yet those farm interest groups don't have the same influence over the Chairman of the Federal Reserve Board, the Secretary of the Treasury, or the responsible authorizing committees of Congress concerned with macroeconomic policy.

This leads to proposals from farm groups such as the one to create a new green exchange rate, i.e. a fixed exchange rate applicable only to agricultural trade transactions as is used in the European Community. Others have proposed interest rate buydowns to reduce the cost of borrowing to farmers. Some economists have suggested designing agricultural policy to blunt the overshooting to which macroeconomic policy subjects agriculture. In fact, as I argued above we may have done that inadvertently in the 1981 farm bill.

This type of request from the farm interest groups can put the agricultural policymaker in a new role. Farm groups may expect the Secretary of Agriculture to be their advocate with the Federal Reserve or the U.S. Treasury and the tax writing committees of Congress.

This role is devolving to the Office of Economics within USDA, where more and more time is spent on assessing the impacts of macroeconomic policy on the farm sector. Unfortunately, the research base on which to draw is less than totally adequate. Attempts to quantify the magnitudes of the linkages between macroeconomic policy and farm sector variables still are few in number and often contradictory.

This is an area where the agricultural economics profession needs to redouble its efforts both in the training of graduate students as well as in research. Far too few agricultural economists trained today have anything close to an adequate training in macroeconomics. This is due partly to a lack of course work, but students with numerous credits in macroeconomics often find little content on the sectoral impacts of monetary and fiscal policy.

Not only does the late 20th century agricultural economics graduate student need to take more macroeconomics courses, but she or he also needs to find more sectoral analysis built into macroeconomics courses by our general economics departments. In addition, students need more training in general equilibrium analysis. In practice, trade theory is the key place where this occurs. But in addition, students need to understand computable general equilibrium modeling--and this involves more than just adding a few equations onto a simultaneous equations model of the agricultural sector. Empirical general equilibrium analysis will be necessary to understand the adjustment processes that lead to changes in the shape of the economy's production possibilities frontier. Once more of our profession is equipped to do this kind of research, we should be in a better position to build up our research capital stock in this area.

PART III

LINKAGES IN DEVELOPING COUNTRIES

8
Some Issues Associated with Exchange Rate Realignments in Developing Countries

G. Edward Schuh

Back in the days of the Bretton Woods fixed exchange rate system (prior to 1973), which was also characterized by relatively modest international capital markets and reasonably stable monetary conditions, exchange rates were of only modest and sporadic interest to most developing countries (and to developed countries as well). Countries that managed their domestic macroeconomic policies reasonably well (such as Mexico) seldom had to change the value of their currencies. Countries that did not manage their macroeconomic policies well were faced with the need for periodic and politically painful devaluations.

In the last fifteen years, however, the exchange rate has surged to the fore in many cases as the key policy issue. Policy dialogue between the World Bank, the IMF, and developing countries focuses almost inevitably on exchange rate issues. Changes in exchange rates are generally integral to restoring equilibrium in external accounts and to providing proper incentives to stagnating economies.

This increased importance of the exchange rate is due in large part to changes in the international economy. These include the shift to a bloc-floating exchange rate regime in 1973; the emergence of a huge, well-integrated international capital market; a significant increase in monetary instability in the international economy; and the rapid growth in international trade during the 1970s, which made

---

G. Edward Schuh, Dean, Humphrey Institute of Public Affairs, University of Minnesota, and former Director, Agriculture and Rural Development, The World Bank, Washington, D.C. The remarks herein in no way constitute official policy of The World Bank.

most economies more open and beyond the reach of domestic economic policies.

In my remarks I have chosen to discuss some issues associated with exchange rate policy in developing countries and their implications for agricultural commodity markets. This is in keeping with the central mission of the Agricultural Trade Research Consortium, which is to promote research on agricultural trade issues. I hope my remarks identify some important and interesting research problems for you.

I want to discuss seven topics: (i) the importance of exchange rate issues; (ii) some issues that arise in adjustment programs; (iii) third-country effects of exchange rate realignments; (iv) the instability problem; (v) stabilization issues ; (vi) income distribution consequences of exchange rate realignments; and (vii) rigidities in the system and their general consequences. Lest there be confusion, let me emphasize that except where noted, I refer in my discussion to the real exchange rate, where this refers to the relative price of tradeables versus nontradeables, or the relative price of tradeables versus labor costs.

IMPORTANCE OF EXCHANGE RATE ISSUES TO AGRICULTURE IN DEVELOPING COUNTRIES

Three developments or situations have brought exchange rate issues to the fore in thinking about agricultural policy and agricultural development problems. The first is that most developing countries have in the past overvalued their currencies for one reason or another. This is significant for agriculture because it is the most well-integrated internationally of any sector of our global economy. Most countries of the world either export or import agricultural commodities, and many do both. For the developing countries, agricultural trade is especially important, and agriculture itself tends to be the largest sector of the economy.

An overvalued currency is an implicit tax on exports and an implicit subsidy on imports. Either way, it has deleterious effects on agriculture. For most developing countries, the overvalued currency is combined with high protection of the industrial sector to siphon resources from agriculture to the industrial sector. Simultaneously, it provides an effective means of keeping food prices low for politically volatile urban consumers.

Policy reforms designed to revitalize stagnating economies thus tend to focus on devaluing domestic currencies so as to help realize a nation's underlying comparative advantage. In the context of agricultural sectoral policies, devaluation is generally seen as the means to provide incentive prices to agriculture, thus facilitating the modernization process and promoting growth processes in agriculture.

Changes in the configuration of the international economy provide the second reason why exchange rate issues have become an increasingly important aspect of agricultural policy. As I have noted elsewhere, the shift to a flexible exchange rate system and the emergence of a well-integrated international capital market put the exchange rate at the center of the process by which changes in monetary and fiscal policy affect the economy. With this configuration of the economy, changes in monetary and fiscal policy impact the economy through changes in the trade sectors (export and import-competing), and the means by which these changes are brought about are by changes in the exchange rate. Agriculture as an export sector and as a sector that competes with imports is thus a sector that has to bear this adjustment. This is a significant change from the era in which exchange rates were fixed and international capital markets were less extensive and less well-integrated.

The third reason why the exchange rate has increased in importance is the emergence of the international debt crisis. The debt crisis is an important byproduct of the petroleum crises of the 1970s, which created an environment in which commercial banks were urged to recycle the petro-dollars in order to keep the international system from collapsing. The bankers recycled with alacrity, and the developing countries were willing borrowers because this enabled them to avoid the exchange rate realignments and the decline in real income which would have been the alternative.

Unfortunately, most of these loans were on very short terms and with flexible interest rates. When the United States changed policy in late 1979 and stopped monetizing its own government debt, the result was a large rise in interest rates and a similar large rise in the value of the dollar. Hence, the developing countries found themselves not only with a large rise in interest rate payments, but an increase in the price of the dollars needed to service the debt. To complicate things even more, U.S. stabilization policies led to disinflation, and commodity prices as usual declined earlier and further than the prices of manufactures and services.

D-day had arrived for those countries that had taken on a large debt. The devaluations that had been deferred by borrowing now became inevitable, even though governments continued to try to avoid the inevitable. Consequently, devaluations have become a big part of conditionality on the part of the World Bank and the IMF.

ISSUES IN ADJUSTMENT

A key component of enabling countries to improve their ability to service their foreign debt is to adjust their domestic economy to a more open or externally-oriented configuration. Fundamentally this means increasing the share of the economy that produces exports, but it also means reducing imports, at least in the short run.

There are three issues I would like to focus on in discussing the adjustment problem. The first is the sources of the increased supply of exports. Opponents of exchange rate realignments tend to be pessimistic about domestic supply response, and thus argue that little can thus be expected from devaluations. The problem with this argument is that it mistakenly assumes a one-to-one relationship between domestic supply and export supply. In point of fact, the export supply is an excess supply function, with the result that a significant part of the adjustment can be expected to come in the quantity demanded domestically for the commodity. Even with no domestic supply response there could still be a significant positive export supply response. Moreover, given that most countries are only marginal exporters, or export only a small proportion of their total production, the elasticity of this supply response can be expected to be relatively large.

Second, there is a general failure to recognize what a devaluation does on the capital account. Devaluation makes domestic assets cheaper in terms of foreign currencies. Consequently, if purchase of those assets, whether they be land or other local investments, is permitted, an inflow of capital can be expected from devaluation. This inflow of foreign capital can facilitate the adjustment process, or it can reduce the need for adjustment per se.

Finally, there is the role of exchange rate policies in the overall capital-markets. The issue here is that fixed exchange rates lead to speculation against future devaluations. The frequent result is large capital outflows that are counterproductive in terms of the larger adjustment problem. The solution to this problem is to put some downside and upside risks in the foreign exchange markets. The best way to do this is by shifting to a flexible exchange rate system.

THIRD COUNTRY EFFECTS OF EXCHANGE RATE REALIGNMENTS

Third country effects of exchange rate realignments are a logical consequence of the present system of a bloc-floating exchange rate regime. These effects have been quite significant in some cases, although their importance does not seem to be sufficiently well recognized by policy makers.

Brazil's experience in the 1970s and early 1980s provides a convenient means of illustrating this problem. As a country which imported approximately 85 percent of its petroleum at that time, the rise in petroleum prices in 1973 literally mandated a significant devaluation of the cruzeiro in order to spread the effects of this shift in external terms of trade throughout the economy and to bring about the adjustments in the domestic economy needed to respond to this change in external conditions.

Brazil at the time was pursuing a crawling peg exchange rate policy designed to keep its real exchange rate constant in purchasing power parity terms. It declined to change from that policy and instead opted to pursue its import-substituting policies a bit stronger as a means to bring about equilibrium in its external accounts.

It turned out that this policy worked reasonably well, even though it may not have been optimal by most standards. During the ensuing years Brazil experienced one of the highest growth rates of any country in the world. But it did so for obvious reasons. The value of the US dollar fell significantly during this period, and since the cruzeiro was tied in large part to the value of the dollar, its value fell relative to those currencies which were changing vis-a-vis the dollar. Moreover, competitive relations vis-a-vis the US remained virtually the same, and the US is a large market for Brazil.

The year 1979 brought another large increase in petroleum prices, and Brazil was faced with the same dilemma it faced in 1973. Its logic was impeccable. It had not devalued in real terms in 1973, so why should it do it now? (To its credit, Brazil did take a maxi-devaluation in 1979, but failed to follow monetary policies that would have brought about the needed changes in domestic terms of trade. The effect on the real exchange rate was thus soon lost.)

It turned out that external conditions also changed in a very significant way. Rather than the US dollar falling once again, the US changed its monetary policies, as noted above, leading to an unprecedented rise in the value of the dollar. Under these circumstances, Brazil's economy was clobbered by its constant real exchange rate policy. The problem was exacerbated, of course, by the enormous rise in real interest rates associated with US monetary policies, and the need to service the debt with dollars that were even more expensive in terms of domestic resources.

As they say in the movies, the rest is history. The important point, of course, is that even had Brazil taken a change in its real exchange rate in 1979, it would still have faced serious adjustment problems. By not moving to take these adjustments immediately, the consequences accumulated to crisis proportions and growth stagnated.

Similar conditions are back of Mexico's problems, although in that case the problem was even more severe and acute due to Mexico having fixed its nominal exchange rate. The consequences of doing this was a rise in the real value of the peso, and an even more severe crisis. This is what has led to the enormous flights of capital from Mexico, and the huge lurches in both the nominal and real values of the peso.

Ironically, countries which pursue such fixed exchange rate policies believe that by so doing they escape adjustments to their economy imposed from abroad. In point of fact, however, the consequence of such policies is to create crisis conditions, the loss of capital, and a concentration of the adjustment in a short period of time. This is probably the worst of all possible worlds.

THE INSTABILITY PROBLEM

An important aspect of today's world is that international financial flows are almost completely dominating foreign exchange markets. In 1984, the last year for which I have data, total international financial flows amounted to US$42 trillion, thus swamping the US$2 trillion in international trade.

Unfortunately, our thinking about exchange rates tends to be dominated by a trade perspective. We tend to think that exchange rates are determined by underlying competitive advantage, and fail to recognize the importance of domestic savings rates, monetary and fiscal policies, and capital flows.

In today's world, capital and financial flows tend to create two kinds of instability in foreign exchange rate markets. First, there is a great deal of short-term volatility; second, there are large, multi-year swings, as exemplified by the fall in the dollar in the 1970s, its unprecedented rise in the first half of the 1980s, and now its multi-year decline again.[1]

I want to focus on the long swings, because they raise a number of important issues. In the first place, what one has is a tendency for capital-flow-driven changes in the exchange rate to mask underlying comparative advantage. From an agricultural standpoint, how does a country position itself vis-a-vis these long swings, especially when many of the investments needed to strengthen agriculture, such as agricultural research, tree crops, and irrigation projects have long gestation periods? The design of development programs under these conditions becomes quite difficult. Equally as important, there is a lot of economic wastage associated with having zigged based on *ex ante* conditions when in *ex post* terms producers should have zagged. Similarly, the kind of flexibility producers and countries need in order to accommodate this world of instability results in a decline in specialization and in turn a loss in the gains that might have resulted from specialization-induced international trade.

This is one of the most poorly understood issues on the international scene. We need to understand it a great deal better if we want to realize the gains from specialization and international trade and thus promote global economic expansion.

STABILIZATION ISSUES

Southern Cone countries of South America--Chile, Uruguay, and Argentina--have used exchange rate policies as the center-piece of their domestic stabilization programs, or more specifically, their attempts to eliminate the unusually high rates of inflation that characterize these economies. Brazil is now following a somewhat similar set of policies.

Let me draw on the Chile experience to illustrate how this policy works. The essence of the policy was what is described as an active crawling-peg exchange rate policy which attempted to draw on rational expectations. The center-piece of the policy was to target the rate of devaluation to exceed the rate of inflation, with large announcement effects. The reason for setting the devaluation to exceed the rate of inflation was to compensate the import-competing sector for tariff-reductions that were an integral part of the stabilization policy. Note that the devaluation is announced as a schedule extending into the future. This pre-announced devaluation was to be the main means of stabilization.

For a couple of years, this policy worked well. GDP grew at a high rate, inflation declined, exports increased--including in agriculture--and the balance of payments crisis disappeared.

But then in 1979 policy makers fixed the exchange rate in nominal terms, at the very time that the monthly rate of inflation was about 2.5 percent per month, substantially above international levels. The result was economic chaos. Chile has since changed its policies and is now operating with essentially a floating exchange rate system. In fact, from a policy standpoint, Chile is now one of the more interesting countries to study because of the innovations in its policy.

The use of an active exchange rate policy as the basis for stabilization policy in both Argentina and Chile eventually encountered difficulties due to developments in capital markets. In both cases, at some stage in the implementation of the policies capital inflows were leading to real appreciations in the value of the domestic currency. Chile, for example, had had two revaluations prior to fixing the value of the currency in real terms.

The effects of these policy experiments on agriculture have to date received very little attention. They deserve more attention on our part.

INCOME DISTRIBUTION CONSEQUENCES OF EXCHANGE RATE REALIGNMENTS

As the most important price in an economy, realignments in exchange rates can be expected to have significant effects on the distribution of income. And they do, often in ways that either make it difficult to bring about the change in exchange rates in the first place, or that cause the realignment to be abandoned shortly after it is implemented.

In the case of agriculture, a key issue is what devaluations do to the prices of food for politically volatile urban consumers. After all, a common reason for over-valuing currencies is precisely to keep the prices of food low to such groups. The consequence of devaluing the currency is to shift the terms of trade in favor of agriculture and thus to raise food prices. This may be either in terms of domestically produced commodities, or in terms of imports.

To the extent that food prices are raised to low-income, disadvantaged groups, well-targeted feeding programs provide at least a partial answer. We can benefit from more design work on such programs, as well as research designed to better understand what the income distribution consequences are.

But food is also a more general wage good. To the extent that the rise in food prices eventually leads to a rise in nominal wages to compensate them, the consequences extend both to the private and public sectors more generally, in the latter case with important fiscal implications. We understand these effects only very poorly at this time.

Finally, we should recognize that low-income groups are not the only groups that benefit from over-valued currencies and thus are harmed by devaluations. Upper and middle income groups are also affected, and in some cases--such as in Brazil--these are the groups that dominate the policy-making process.

My main point is to emphasize how little we understand in an empirical way about the income distribution consequences of exchange rate realignments, and how important an understanding of them is to bringing about needed policy reforms. I know of no issue more important in bringing about policy reform in most developing countries. If we don't find ways to deal with these income distribution consequences, we simply will not bring about the reforms that we so desperately need.

EXCHANGE RATE RIGIDITIES

The current bloc-floating exchange rate system has rigidities in it that impose negative externalities on all participants in the system. Some of these involve formal arrangements, such as the European monetary system of fixed exchange rates and the French franc bloc in Africa. But others are less formal, and involve the general tendency for developing countries to fix the nominal or real value of their currencies to the value of the U.S. dollar, the British pound, the French franc, etc.

A return to a unified fixed exchange rate system is probably no longer in the cards. Capital markets have become too extensive and too volatile to make that possible, and the likelihood of coordinated macroeconomic policies among countries with the major currencies is in my judgement quite slim. Unfortunately, the present half-way house of some countries pursuing fixed exchange rate policies and others pursuing flexible exchange rate policies creates an inordinate amount of instability. Fixed exchange rates cause distortions to build up and adjustments to be delayed. They also induce large capital flows among countries. The result is to impose large shocks on the international system, and especially on the economies of those countries which pursue floating exchange rate policies.

The international monetary system is in bad need of reform, as I have argued elsewhere. Modest steps to that end are now emerging. In the interim, many developing countries can benefit both themselves and the international economy by shifting to a flexible exchange rate system. Again, this shift could be facilitated by research which seeks to understand the consequences of such changes and which provides guidance to policy makers on how to bring about changes in existing policies.

CONCLUDING COMMENTS

The international economy has changed dramatically these last 20 years, and the role of exchange rates has probably changed as much as anything. These changes have pervasive implications for agriculture and for agricultural commodity, trade, and development policy. Until we better understand this new system, we will continue to make major mistakes in policy. The research tasks before us are enormous.

NOTES

1. These large capital flows, of course, have tended to make interest rates more stable, the logical consequences of having a well-integrated international capital market. But that is why changes in the exchange rate have become the key issue.

9
Financial Constraints to Trade and Growth: Crisis and Aftermath

Mathew Shane and David Stallings

INTRODUCTION

The debt crisis, which began with the international debt-repayment problems of Poland in 1981 followed by those in Mexico, Brazil, and Argentina in 1982, has proven to be a far more serious threat to the world economy than anyone anticipated (Shane and Stallings, 1984a). The problem that was initially perceived as a threat to the stability of the international financial system has turned out to be a more binding and intractable constraint on international trade and development.

Using 1970-85 data from 79 developing countries, we evaluated the course of events which led to the debt crisis, the adjustments which have taken place since 1982, and the prospects for renewed growth under the existing debt resolution strategy. Some observers, in the early days of the crisis, assumed that a 3-year adjustment period would be sufficient to overcome any short-term disequilibrium in the world payments system (International Monetary Fund, 1984; Shane and Stallings, 1984a). This adjustment period would soon be followed, they argued, by renewed growth based on revised and strengthened trade alignments. To date, however, no evidence of renewed sustainable growth in the problem debtor countries has surfaced. Furthermore, the constraints on the most debt-affected countries may very well be retarding the entire world growth and trade system.

Mathew Shane and David Stallings are economists in the Economic Research Service, U.S. Department of Agriculture.

Many lenders significantly reduced the amount of credit available to developing countries in the aftermath of the debt crisis. This withdrawal accelerated into 1985-86 and continued in 1987. Developing countries received some $57 billion in credit in 1978. Credit availability declined by almost $90 billion per year during 1982-85, so that repayments exceeded new lending by more than $30 billion per year.[1] Furthermore, this imbalance has led to steep declines in gross capital formation and a dramatic falloff in per capita income growth.

The adjustment to the debt crisis, therefore, did not lead to renewed growth in trade and development but instead to declining trade worldwide and stagnating per capita incomes. . Rescheduling debt prolongs the problem. By treating the debt problem as merely one of liquidity, this solution actually may have lead to a situation in which the global effects of the cure are worse than the failure of some countries to meet their international debt servicing obligations. Dramatically different solutions for overcoming the debt crisis (such as forgiving or writing down some portion of the debt incurred by the most severely indebted countries) are more likely to place developing countries on a growth path consistent with that observed prior to 1982.

We found that the negative effects of the debt crisis on trade and development may well be greater than the potential costs of forgiving some portion of some countries' international debt. The cumulative effect of changes in policy for a set of countries which individually are relatively small parts of the world trading system can add up to a total effect on world trade that is quite substantial. Until 1983, the middle income debtor countries were the fastest growing segment of the global economy.

THE ANTECEDENTS

The current world debt problem had its roots in the rapid growth and development of the 1960s and early 1970s when credit was readily available and inexpensive. That long period of sustained world growth created excess demands for natural resources. That excess demand for resources, most notably petroleum, provided the conditions under which the Organization of Petroleum Exporting Countries (OPEC) could be formed and become an effective force for monopolizing world petroleum trade.

The fourfold increase in petroleum prices initiated by OPEC in 1973-74 substantially shocked the world economy. The principal shortrun effect was to create, for most trading countries, a balance-of-trade disequilibrium. The high-income oil exporting countries generated significant trade surpluses. At the same time the oil importing countries generated balance-of-payment deficits. The longer term effect of the oil price increase was significant debt accumulation by developing countries, setting the stage for the current world debt problem.

The industrial countries employed easy monetary policies both before and after the first oil shock, which moderated the decline in real income and permitted continued economic growth in developing countries. The change in trade flows and expansionary monetary policies in the member nations of the Organization for Economic Cooperation and Development (OECD) generated large amounts of money previously unavailable to the international financial system. International bankers recycled this liquidity in the form of "petrodollar" deposits by beginning a massive lending program focused primarily on middle-income developing countries. These bankers anticipated high returns on investments and assumed that a country guarantee was adequate provision against repayment defaults. The bankers did not ask if the funds were being invested in such a way that a stream of foreign exchange earnings would be forthcoming to repay the loans. As we argue later, the characteristics of the world economy during the latter half of the 1970s would have made such an investigation superfluous.

The world economy weathered the first oil crisis without much apparent difficulty. Initial debt levels were low enough that accumulation did not overly burden the world payments system. Furthermore, the infusion of large amounts of international capital into the world economy generated an international expansion led by export growth. For all non-OPEC developing countries, the total dollar value of exports was 2.5 times greater in 1980 than in 1975. Furthermore, annual real growth in gross domestic product (GDP) for all developing countries averaged 5 percent during this period.

The oil price rise of 1973-74 set the stage for the large debt accumulation, and the second oil shock of 1979-80 set the stage for the world recession of 1980-83. The latter petroleum price increase was more significant than the first because of the large debt that had accumulated and the far different policy responses of the industrial nations. The reaction to the 1979-80 increase was for the major industrial countries to simultaneously restrict available credit.

The resource-driven inflation that was initiated by the 1973-74 oil price increase and accommodated by expansionary monetary policies proved unacceptable to the industrial countries. The rapid and uncontrolled rises in resource costs were significantly reducing real manufacturing profits, eroding confidence in the future, and lowering investment. Only traditional measures could deal with the anticipated inflation.[2] The sudden decline in monetary growth sharply slowed the world economy, raised real interest rates, and made the debt a burden. The effect of the policy responses of the developed countries to the second oil shock triggered the current repayment problems.

THE MACROECONOMIC POLICY ENVIRONMENT

If the oil price shocks of the 1970s led to changes in the monetary policies of the industrial countries, the growing world integration of capital markets transmitted the changes from lenders to borrowers and magnified the growth of international credit availability.

When exchange rates are flexible, monetary policies tend to initially affect the domestic economy by changing interest and exchange rates. Expansionary monetary policies drive domestic real interest rates below international rates and thus create external incentives for domestic money holders to place or create assets overseas, as happened in the United States during the 1970s.

Short-term interest rates on dollar-denominated deposits in London were consistently above available rates in the United States during the 1970s (figure 1). Clearly, rational actors in the U.S. banking system would tend to place increasing amounts of "idle" deposits overseas. As a result, given no reserve requirements, the growth rate of world overseas assets and liabilities (about 80 percent of which were in U.S. dollars) was significantly higher than that of U.S. M1 (the total of all U.S. currency and all checking deposits) during the 1970s and lower during the 1980s.[3]

Overseas bank assets are a much more effective measure of world liquidity[4] than the simple total of national money stocks (figure 2). First, such assets are the base used for much of world trade and financial flows. Second, this measure of money, in the context of the international economy, better explains the essential results of what, given the evidence below, indicates the occurrence of a worldwide monetary shock.[5]

Movements in Overseas Bank Assets

Overseas bank assets grew rapidly and continuously through the 1970s, increasing at an average annual rate exceeding 27 percent between 1973 and 1981 (figure 2). However, the increase declined abruptly to less than 8 percent in 1982, followed by 2 years of less than 5-percent growth. The slowdown matched the decline in the rate of debt accumulation by the developing world (figure 3). Only in the last two quarters of 1985 and through 1986 did world liquidity expand in a manner similar to percentage increases observed in the 1970s.

Several factors seem to dominate the slowdown in world money growth. First, deregulation of the U.S. domestic banking sector removed one of the chief incentives for overseas deposits by U.S. investors and U.S.-owned international banks. Second, the balance-of-payments adjustments to debt constraints reduced the demand for overall international liquidity. Third, the fall in income in the developing countries led to a slowdown in world trade and lessened the demand for money for international transactions. Finally, the domestic money demand function in the United States significantly shifted in 1981-82, sharply increasing the aggregate demand for money.[6] The increased desire to hold money in the United States would, in and of itself, reduce the supply of dollars formerly available to the world trading community and lower the overall quantity of overseas bank assets.

The key in examining the monetary aspects of the debt crisis comes in considering the transmission mechanism by which the integration of world capital markets (facilitated, to a large degree, by the growth in offshore banking centers) significantly exaggerated the incidence of debt accumulation.

The Transmission Mechanism

Although the major developed countries moved to a flexible exchange rate system in 1973, the developing countries have, for the most part, maintained fixed exchange rate regimes aligned with major currencies. Because these countries essentially respond to changes in world monetary conditions, we can analyze their reactions to changes in the growth of offshore bank liabilities.

An increase in offshore money will, by depressing offshore interest rates, lead to capital inflows, until domestic and overseas real interest rates equalize. Foreign exchange reserves will increase, and the domestic money stock will rise as foreign currency is traded for local currency.

However, during the 1970s, many developing countries chose to allow domestic money to rise more slowly than world liquidity. This "sterilization" resulted in rapid reserve accumulation and price distortions between nontraded and traded goods.[7] Interest rates also became "unbalanced", as domestic real rates of return would remain higher than those prevailing in world markets. Real exchange rates would appreciated.[8] Debt accumulation under these circumstances is certainly rational: borrow at low rates, and repay with earnings that outpace interest due. The willing financing of current account deficits was not viewed as a disequilibrium phenomenon.

The rapid increase in world money during the 1970s not surprisingly resulted in rapid debt accumulation. The situation changed drastically, however, when the easy money times of the 1970s were abruptly transformed into the much different international financial environment of the 1980s.

The money shock of the early 1980s produced a dramatic reversal in the direction of the real interest rate advantage. The sterilization of reserve outflows suddenly resulted in more rapid inflation. Real depreciation was the implicit policy response. Lower domestic returns now had to support the higher real repayment schedules contracted in the early 1980s. Loans assumed at variable rates would necessarily prove particularly difficult to service. Those countries that undertook monetary sterilization found real repayments growing faster than real income.

External reserves, real interest rates, trade flows, and price changes all reflect the expected outcome of the sterilization policies followed by the most severely affected debtor nations.

Reserve Flows

For the 79 countries, the 1970s saw the dollar value of all reserves other than gold rise at annual rates exceeding 20 percent (figure 4), before plummeting during 1980-83. Reserves did not return to the level of 1979 until 1984. The reserve buildup during 1984 may have acted as an additional constraint to the adjustment of the most debt-affected countries by diverting resources that could have been used to repay debt or to purchase needed imports.[9]

Sub-Saharan Africa remains in the most precarious position (figure 5), with total reserves at the end of 1985 barely one-third the level of 1980. (See Appendix A for the countries included in each category). Other categories whose reserve positions have not yet returned to the levels of 1980 are South Asia, Latin America, North Africa and the Middle East, low- and middle income countries, oil exporters, major borrowers, and debt-affected major borrowers. Southeast Asia and Northeast Asia have accumulated reserves over the period. The domination of the Asian countries in the major market group also reflects their small increase in reserves over the period. All categories (except North Africa and the Middle East) substantially increased their reserves during 1984.

The change in reserves mirrored movements in the current account balance until 1980, when the sudden increase in current account deficits reflected a sharp decrease in reserves (figure 6). The movement in reserves is, in fact, more closely related to changes in world liquidity. The money shocks in 1981-83 forced a drawdown in foreign exchange. The exceptions are the countries of Northeast Asia, where reserves have accumulated since 1979, regardless of the external position. Current account surpluses plus reserve accumulation place this region in particularly good position for adjustment to any future external shock.

Reserve/import and reserve/export ratios also altered significantly from 1973-80 to 1981-85. The former period had reserve/export ratios for all countries at 27-31 percent (figure 7). The average for the 1980s, to date, is below 20 percent. Northeast Asia is again the exception; its ratio has increased during the 1980s (figure 8). The Latin American nations showed an especially sharp decline in the reserve/import ratio between 1979 and 1982, before the slight rebuilding in 1983 and 1984. The rise in the reserve/import ratios in 1983/84 also partially reflects declining imports.

Changes in Interest Rates

Market interest rates have grown in importance in loan repayments, particularly since 1978-79. Loans extended at variable interest rates, with premiums at fixed points above the U.S. prime rate or the London Interbank Offered Rate (LIBOR), became popular during the late 1970s.

Real interest rates incorporating price changes provide a measure of the current opportunity cost of debt repayment. The U.S. real interest rate is typically derived by subtracting current inflation (or some series of recent measures that reflect expected inflation) from nominal interest rates. The appropriate measure for debtor countries is the interest rate adjusted for changes in export prices. If export prices rise faster than contracted interest rates, the real rate is negative.

The effect of the rapid increase in money during the 1970s is clearly seen when compared with the real interest rates faced by the developing countries. That decade was dominated by price increases far exceeding nominal interest rates (figure 9). The lowest real interest rates were those experienced by the oil exporters and the Middle East and North Africa countries. The nations of Sub-Saharan Africa faced the least favorable situation.

The phenomenon of negative real interest rates is fully in keeping with the transmission mechanism described above. Creditors received the benefit of higher nominal returns in their own currencies, and debtors were able to capitalize negative external real rates into domestic investment opportunities. Moreover, even an investment which yielded negative real returns at home could have been higher than the negative repayment rates and, when viewed externally, still be relatively profitable. Further, even a constant volume of exports would meet contracted obligations. There existed no incentive, in the aggregate, to raise the production potential of export industries.

The situation of the 1970s quickly reversed itself in the 1980s. Nominal long- and short-term interest rates on dollar loans rose sharply beginning in 1978 as rising inflation began to add premiums to the cost of borrowing. Not until 1981, however, did price increases for exports from the developing world fall below nominal interest rates, and the real rate increased sharply. Despite the decline in short-term rates in 1983-85, real interest rates facing all developing countries remained above 10 percent and were higher in 1985 than in 1984 for 13 of 15 country groupings, the exceptions being Yugoslavia and North Africa and the Middle East.

The highest real interest rates are faced by Latin America, Southeast Asia, and the debt-affected major borrowers. None of the country groupings have real "long-term" repayment rates below 10 percent.

Exchange Rate Movements

The real depreciation of the U.S. dollar against the aggregate of currencies during 1972-78 has been completely reversed during the 1980s into 1986. The real value of the dollar has risen by more than 50 percent between 1981 and 1985, against the currencies of the 79 countries in our study. In 1985, the U.S. dollar was at its highest level since the collapse of the Bretton Woods system in 1973.[10]

The currencies of South Asia have been continually devalued since 1974, while those of Northeast Asia have declined by 50 percent since 1979 into late 1985. Latin American currencies were devalued some 23 percent in 1982 alone and by 45 percent in 1985 from 1981. Major U.S. agricultural markets have continued to allow their currencies to depreciate at an accelerating rate between 1979 and 1985 (figure 10).

Exchange rates are used as policy instruments by most developing countries. Only a few of the currencies of the 79 countries we studied had their values determined in free markets as of 1985 (for example, the Dominican Republic's peso and Costa Rica's colon now have their values determined by domestic banks). Most adjustments are at infrequent intervals and tend to be doubly disruptive when anticipated. Reserves, for example, may be depleted when individuals expect a devaluation, or foreign exchange may be rationed. Depreciation occurs when financing is unavailable to cover current account deficits, repayments, and reserve accumulation. In 1982, credit flows declined as world liquidity contracted, reserves began to disappear, and developing countries initiated significant real devaluations.

The severity and suddenness with which Latin America and the other debt-affected nations devalued their currencies dramatically demonstrates the seriousness and sharpness of the shift in the international monetary environment, as well as pointing to poor domestic resource allocation between nontraded and traded goods during the 1970s. Domestic adjustments, particularly in accelerating rates of inflation, were severe. The price for overvaluation was reduced imports of goods and services that contributed to economic growth. After subjecting their economies to sudden consumer price index (CPI) increases in the 1980s, both Brazil and Argentina have changed currencies. Brazil fixed its exchange rate, temporarily at least, at 13.8 cruzados to the dollar in March of 1986 and vowed to follow passive monetary policies rather than sterilization. Pressure on net export earnings forced significant real devaluations, however, beginning in August 1986.

The debtor nations were therefore caught in a difficult situation. The principal on loans that had been falling in real value began to rise at an accelerating rate. The declining real repayments so evident and welcome during the 1970s also began, in 1981/82, to rise in real value.

Consumer Prices

One of the most telling of adjustment indicators in the domestic economy is the inflation rate. Measured as the change in consumer prices, the rate that general prices increased accelerated in all country groupings (figure 11) except Northeast Asia.

The most dramatic single country case is that of Bolivia, which in 1985 had the highest inflation rate in the world, 100,000 percent over 1984. Because Bolivia had such a large weighted inflation change, we present the CPI patterns with and without Bolivia (figure 11).[11] Particularly large rises in the inflation rate during 1980-85 occurred in Latin America, where the increase ranged from 45 percent to 484 percent per year.

The above situation sharply contrasts with the experience of the Asian regions, all of which had declining inflation rates between 1980 and 1985. Consumer prices in Northeast Asia are now increasing at only 2 percent per year, 240 times less than in Latin America.

Rapid inflation is an economy-deadening phenomenon in countries with limited (or negative) access to world capital markets. Some of its more ravaging nonneutral effects are the elimination of private saving, curtailment of long-term contracts, capital flight, and the virtual end of domestic investment in new productive capacity. This depressing phenomenon is most evident in the gross capital formation in the countries with the highest inflation rates (figure 12). As inflation accelerates, the share of GDP taken by capital formation falls.

Commodity Price Adjustments

The price adjustments that have taken place in the world trading sector reflect the influence of the changing growth rate in world liquidity and its transmission to developing countries. The real appreciations of the developing countries' currencies during the 1970s and general raw material shortages contributed to the price increases of the period. Those same factors were reversed in the 1980s as export promotion (real devaluation) policies accompanied excess stocks of primary, raw commodities important to trade from poorer countries. Price changes directly reflect the sharply different exchange rate, interest rate, and monetary environment of the 1980s compared with the 1970s.

The basic interaction between exchange rates and prices is one of the most direct in economics. When the value of foreign currency rises, individuals must give up more of their local currencies to obtain the same amount of foreign currency as before. All goods sold in units denominated in dollars, for example, will appear to rise in price. The supply curve appears to offer less at every price, thus, reducing supply. The seller must accept a lower dollar price in order to sell the same amount; the demand curve will appear to rotate clockwise. A depreciating dollar would have the reverse effect.

Factors other than exchange rates also affect the amount people sell and the quantities that others are willing to purchase. Variable weather, cartels, and changing market conditions have had profound effects on the supply of a variety of internationally traded goods. Wide swings in the growth rate in world income over the past 20 years have also significantly affected the ability of purchasers, both actual and potential, to buy products offered for sale in world markets.

Between 1973 and 1980, the value of the dollar declined by 35 percent against all currencies. During the same period, prices, as measured by export unit values, more than doubled (figure 13). During the 1980s, the situation reversed that of the previous 8 years. The dollar rose by 40 percent between 1980 and 1985, while export unit values have fallen by 15 percent. Many individual commodities and commodity indexes have fallen by far greater amounts during the same period, however. The all primary commodity index[12] has declined 25 percent since 1980; raw food commodities such as grains and fruits have also fallen 25 percent. The index of all metals has dropped 30 percent, copper is down by 35 percent, and tin has fallen 28 percent in price. For Brazil, the dollar export price of sugar has fallen by more than 60 percent.

The last time price declines were as uniform as during the 1980s was during the Great Depression years of the 1930s. For many of the most debt-constrained countries, the comparison is apt.

Interest rates, in addition to their role in capital flows and exchange rate determination, also exert considerable influence of their own over prices. Most production and sales are protected by some sort of inventory "buffer" which smooths uneven cycles in supply and demand. Interest rates are crucial to the size of these stocks. High interest rates make holding inventories expensive in two ways. First, the cost of borrowing to finance carryover increases. Second, the present value of such holdings declines as real interest rates increase. Both of these factors encourage reduced inventories. The desire to lower inventories shifts supply curves and tends to lower prices.

Money growth also affects prices through the ways in which people spend increased income, although that influence occurs after a greater lag than the effect on interest rates. With increased money, people find themselves with larger balances than they want to hold or to save. In trying to convert money to other assets or goods, prices will rise, as aggregate demand increases in relation to aggregate supply. The reverse will be the case when money (or its growth rate) is reduced.

The rate at which money grows affects interest rates in ways that enhance the price effects noted above. A decrease in money growth may raise interest rates by contracting the supply of credit. The result will be to reinforce the price effect of a slower increase in money. Current prices in competitive environments contain all the information reflecting the monetary shock, interest rate, and exchange rate changes of the 1980s.

Trade Patterns

The current account balance is closely related to the flow of credit to the developing countries in the 1970s. The availability of that credit during the 1970s permitted the widening of current account deficits through 1981. Similarly, when credit was curtailed, developing countries had to reduce imports and promote exports.

The current account deficit for all developing countries reached $153 billion in 1981, declined to $60 billion in 1985, and remains concentrated in North Africa and the Middle East and South Asia. Northeast Asia is the only developing country grouping that maintains a surplus. Current account deficits have dropped the most in absolute terms in Latin America (figure 14), the upper middle-income countries, the debt-affected countries, major borrowers, and major U.S. agricultural markets.

Between 1981 and 1985, the total nominal dollar value of exports and other service inflows (excluding unrequited transfers) has remained virtually unchanged for all countries, declining slightly from 1984 into 1985. The total exports of all major borrowers, debt-affected major borrowers, Latin America, Sub-Saharan Africa, North Africa and the Middle East, and low- and middle-income countries have actually declined from 1981. Only the Asian regions and the upper middle-income countries have made significant export gains (figure 15). Major U.S. agricultural markets have seen their exports stagnate.

Most countries have reduced their current account deficits by reducing imports. Total imports have declined by nearly $100 billion since 1981 for all 79 countries. Only the Asian regions have shown an increase over the period. Sub-Saharan Africa has cut imports by more than 30 percent, while Latin America, the oil exporters, debt-affected major borrowers, and North Africa and the Middle East are also down by over 25 percent. The largest absolute declines during 1981-85 were in Latin America (down $50 billion, from $180 billion to $130 billion), North Africa (down $40 billion, from $137 billion to $97 billion), oil exporters (down $75 billion, from $258 billion to $183 billion), and debt-affected major borrowers (down $61 billion, from $174 billion to $113 billion).

Merchandise import volume has fallen sharply during the 1980s; Latin America alone has curtailed imports by over 30 percent since 1981 (figure 16), with even greater cuts by the debt-affected major borrowers. The Asian regions, where imports have actually risen, are exceptions (figure 17). Merchandise import levels for the major U.S. agricultural markets declined only slightly during 1981-85.

The decline in prices implies that the volume of exports has increased for all countries since 1981. Merchandise exports have increased to levels 20 percent higher than in 1981 for all 79 countries. However, this figure reflects an increase of only 5 percent over 1979, and virtually no change between 1984 and 1985.

Merchandise exports, expressed in 1980 dollars, have actually fallen from 1979 levels for Sub-Saharan Africa, North Africa and the Middle East, and oil exporters (figure 18). The largest increases have been in the Asian regions, with Northeast Asia having the largest gain (figure 19). Export volume declined in 1985 from 1984 for the oil exporters, debt-affected major borrowers, major borrowers, middle-income countries, North Africa and the Middle East, and Sub-Saharan Africa.

Some of the "improvement" in merchandise trade has, however, been moderated by continued deficits in the services balance, in both nominal and real terms. Current dollar estimates show a slightly reduced services deficit from 1981 to 1985, mostly concentrated in North Africa and the Middle East and oil exporters, the result of fewer oil field jobs for imported workers.

The nominal services balance for most other country groupings has generally stagnated. The real service balance, however, indicates no change for all countries and a worsening for Latin America and the debt-affected major borrowers (figure 20), reflecting the negative net transfers of the 1980s, discussed in the following section.

THE DEBT PROBLEM

The pattern of international debt reschedulings since 1956 indicates the serious misalignment between payment commitments and the ability of countries to service their debts (figure 21). During 1956-75, only 11 countries were involved in debt negotiation and reschedulings. The total amount rescheduled was only slightly more than $8 billion.[13] Between 1976 and 1980, 11 countries renegotiated $13.5 billion in debt.[14] Although the dollar amount increased, whether the reschedulings posed a serious threat to either the world financial or trading system is debatable. However, between 1981 and 1983, 25 countries rescheduled $55 billion.[15] Clearly, the magnitude of the debt at risk began to threaten the international financial system. Although reschedulings declined significantly in 1984, with 18 countries renegotiating almost $13 billion of debt,[16] the number of countries involved in 1985 (24) and amount of reschedulings ($93 billion)[17] indicate that debt repayment is still very much a problem.

Another aspect of the potential problem of rescheduling is the degree to which they involved commercial, rather than official, debt. All renegotiations and reschedulings before 1976 involved official debt. Since 1981, however, more than 90 percent of the dollar amount involves commercial bank debt. The exposure of large U.S. commercial banks to the debt of oil-importing developing countries provides one measure of the potential seriousness of default on bank solvency (table 9.1). During 1980-85, loans to these developing countries far exceeded total bank capital of the largest U.S. banks. However, the peak year of exposure was 1982 where potential claims were twice bank capital.

Since 1982, the ratio has fallen to below 150 percent, a rate below that of 1980. Because of the recent pattern of reduced exposure and the seeming unwillingness of commercial banks to further lend to developing countries, except under duress, the threat to commercial institutions will be further reduced over time.

PATTERNS OF DEBT ACCUMULATION, COMPOSITION, AND RATIOS

Total debt for 79 developing countries reached approximately. $1,100 billion in 1987. This total is up from $790 billion in 1984 and

TABLE 9.1
U.S. Bank Loans to Oil-Importing Developing Countries

	Largest 24 Banks			Other U.S. Banks		
Year	Total Claims	Capital	Claims as Share of Capital	Total Claims	Capital	Claims as Share of Capital
	--Billion Dollars--		Percent	--Billion Dollars--		Percent
1980	54.4	33.8	161	11.8	19.6	60
1981	67.0	36.5	184	15.3	23.2	66
1982	79.3	39.8	199	19.3	26.4	73
1983	84.6	44.1	192	19.1	30.5	63
1984	89.0	49.7	179	18.8	35.0	54
1985	86.1	58.8	146	16.7	40.0	42

Source: Institute for International Economics.

$760 billion in 1983. A more inclusive measure of developing country debt which incorporates Eastern European and Asian centrally planned countries would bring the estimated total to approximately $1,250 billion (Shane and Stallings, 1984a). The composition of debt noticeably moved toward private short-term debt during 1973-82, but lending has shifted away from short-term credit since then. This reshuffling back toward longer term obligations and away from short-term credit has had the positive effect of reducing debt service payments and thereby reducing repayment pressure.

The Debt Composition

The composition of debt varies from region to region and across economic categories (Shane and Stallings, 1984a; and ERS estimates). Northeast Asia (figure 22) has used the highest degree of short-term credit as a proportion of total debt, while the low-income economies and South Asia have the highest level of official credit and the lowest level of short-term credit (figure 23). Latin America has the lowest relative level of official credit (figure 24).

The geographic distribution of total debt changed substantially during 1973-85 (figure 25). Latin America, Southeast Asia, and Northeast Asia have seen the fastest growth in debt, while North Africa and the Middle East, Sub-Saharan Africa, and South Asia have reduced their shares in the total. However, the geographic distribution has been virtually constant since 1982. The distribution across income classes has closely followed that of the regions (figure 26). The upper middle-income countries have tended to raise their share, the middle-income countries have tended to hold the same position, and the lower income countries have reduced their portion. Again, the relative shares of debt have remained stable since 1982.

The Growth of Debt

The annual growth rate of debt exceeded 20 percent during 1973-81 for all developing countries, but there has been a clear secular decline since 1978 (figure 27). This pattern is similar to that displayed by the shares of medium- and long-term debt. The pattern for short-term debt is similar but more pronounced; short-term debt grew by more than 30 percent in the earlier period and actually declined in 1983-84.

The growth rate of debt displays regional differences. Northeast and Southeast Asia have higher rates of debt accumulation than does South Asia, reflecting greater access to commercial markets. But the difference tends to narrow at the end of the period. Similarly, the upper middle- and middle-income countries had a higher growth rate of debt than did the low-income countries. This difference also tended to narrow in the 1980s. With regard to the growth in short-term debt, the figure for the debt-affected major borrowers grew at a much higher rate than major borrowers and the average for the 79 developing countries. Similarly, Latin American short-term debt grew at a much higher rate before 1982. This situation is certainly one of the symptoms suggesting the payments difficulties that these two groups encountered during 1982-85.

The Northeast and Southeast Asian countries had among the highest growth rates of debt over the 1973-83 period, but only the Philippines, of the East Asian groups, has experienced debt payment difficulties. This situation strongly suggests that rapid accumulation of debt by itself was a necessary but not sufficient condition for the subsequent debt servicing problems in the 1980s. If credit is used to make investments which generate a stream of foreign earnings in excess of payment requirements, then even large debts can be serviced. If the credit is used to expand consumption or for investments with either lower rates of return in foreign earnings than restitution due or a pattern of returns which does not match repayments, then payment difficulties will arise. Clearly, the policy of currency overvaluation and real sterilization could easily produce this outcome. The radical change in liquidity growth and availability, plus that of the world trade environment from 1979-82 severely affected the returns to those investments that were made in the late 1970s.

The Withdrawal of Credit

The withdrawal of credit to developing countries, indicated by the declines in the growth of debt, is magnified when one considers the net flows of credits (referred to as net transfers) which went to developing countries during 1973-85.[18] Between 1974 and 1982, the cumulative net transfers to developing countries equaled about $200 billion (figure 28). In 1978, net transfers peaked at $57 billion.

Starting in 1983 and continuing through 1985, net transfers to developing countries were negative, implying debt service payments were greater than incoming new credit. During 1983-85, there was an outflow of about $76 billion, with 1984 alone accounting for almost $40 billion. Although the absolute level of net outflows in 1985 marginally improved, negative net transfers still averaged over $30 billion during 1983-86.

The above situation is best placed in perspective when considering a 1985 proposal by the U.S. Secretary of the Treasury. At that time, the suggestion was for a 3-year goal of increased funding to developing countries of less than $30 billion, which would result in an average of only a $10-billion improvement in the net transfer position of those countries. His plan would have had to be three times as large to achieve even a zero net transfer position of the developing countries had it been implemented in 1985 alone.

The average difference between the 1974-82 period and the 1983-85 period was $50 billion with the peak difference being almost $100 billion, comparing 1978 with 1984. Viewed from any perspective, that change was substantial and one which had to dampen international trade through the loss of available foreign exchange. The decline in the imports of goods and nonfactor services of the developing countries from a peak of just under $500 billion in 1981 to the estimated 1985 value of $410 billion mirrored this change simply because, given the world trading environment, virtually all of the balance-of-payments adjustment had to come from decreasing exports.

The overall pattern of inflow followed by outflow was pervasive in all categories, but the extremes were dominated by a few groups. Thus, U.S. agricultural market countries mirrored the overall pattern closely, as did Latin America (figure 29). The upper middle-income pattern also closely followed that for all countries while the low-income countries made up largely of South Asian and Sub-Saharan African countries showed a much more stable, although declining, pattern (but without the negative net transfers in the latter part of the period).

The Need for Adjustment

The withdrawal of credit from developing countries required substantial balance-of-payments adjustment.[19] We can calculate this adjustment by computing the change in net exports of goods and nonfactor services required to meet at least interest payments on the debt. Taking this calculation as a ratio of exports yields the net adjustment rate.[20] The pattern of 1973-81 was very different from that of 1981-85. In 1973 and 1974, the years of the first oil shock, the net adjustment rate for all developing countries was less than 3 percent. This rate rose to more than 20 percent in 1975, dropping to just over 15 percent during 1976-80. In the peak year of 1981, concurrent with or directly after the change in the growth rate in world liquidity, the adjustment rate rose to more than 35 percent. In 1981-84, the adjustment rate dropped to just over 10 percent.

Unlike some of the other patterns, there are wide differences in the degree to which countries have undertaken the needed adjustment by lowering imports, switching exports, or both. The oil exporters had a pattern, consistent with relatively large current account surpluses, which differs substantially from that of all developing countries in the period 1973-80. No (or negative) adjustment was the rule. But, after 1981, the pattern mirrored that of all developing countries quite closely.

The upper middle-income and middle-income countries had a pattern which closely followed that of all developing countries, except that the upper middle-income countries required a low degree of adjustment throughout most of the period. The low-income countries, on the contrary, had a pattern of increasing need for adjustment, averaging over 70 percent since 1980.

The debt-affected major borrower countries showed more extreme fluctuations than the pattern of all developing countries and even have an adjustment in excess of requirements in 1984 and 1985 (figure 30). Southeast and Northeast Asian countries had relatively stable adjustment patterns while South Asia had a divergent pattern as did the low-income countries of which it is a major part (figure 31). Latin America and, to some degree, Sub-Saharan Africa had patterns of increasing need for adjustment followed by a substantial correction since 1981 (figure 32). North Africa and the Middle East had a pattern indicating increasing need for adjustment in the 1980s compared with the 1970s (figure 32).

Debt Ratios

One common measure of the burden of international debt is debt service as a percentage of exports of goods and nonfactor services. For all developing countries, there was a 250-percent increase in this ratio between the low of 12 percent in 1974 and the high point of 29 percent in 1982. However, throughout that period there were positive net transfers so that this increase in the debt service ratio, was a potential but not an actual burden; new borrowings exceeded debt service payments.[21] Beginning in 1982, the debt service payments became a burden, and Mexico became the first to negotiate reschedulings of its debts. Yet even during 1983-85, net debt service payments amounted to less than interest payments. The debt service ratio declined between 1982 and 1985, most notably between 1982 and 1983. However, even at the reduced rate of 1983-85, one out of four export dollars was going for debt service payments.

The most severely affected country groups show the largest absolute decline in the debt service ratio: upper middle-income countries, debt-affected major borrowers, and Latin America (figure 33). The middle- and low-income countries, South Asia and Southeast Asia (figure 34), and the poorest African countries (figure 33) show continuing increases.

Although the debt service ratio indicates the current debt burden, this measure depends critically on payment terms, amount of new borrowings, and reschedulings. The rescheduling of debt lowers the current debt burden, as measured by the debt service ratio, but also transfers the burden to the future. The debt/export ratio and debt/GDP ratio are two measures of the cost of repaying debt. The former indicates the amount of exports to be forgone for debt repayments and the second the amount of domestic income.

The debt/export and debt/GDP ratios do not show the favorable declines which the debt service ratio indicated. Overall, after the ratios doubled between 1974 and 1982, they leveled somewhat between 1982 and 1985 (figures 35-36). The debt/export and debt/GDP ratios, in every case, were higher in 1985 than in 1982.[22]

Savings from Concessionary Interest

One of the factors which can mitigate the debt problem for developing countries is the degree to which credit is given on concessional terms. One measure of this relief, although an imperfect one, is the degree to which the average interest rate which a country actually pays on its debt is different from the commercial rate. As a proxy for this measure, we computed the savings generated by the difference between the average rates on long- and medium-term debt and on short-term debt.[23] The savings from concessionary interest were modest through 1977 for all categories of countries. However, these savings became substantial starting in 1978 and rose rapidly to 1981, the year of maximum nominal commercial rates. In 1981, concessionary savings for all developing countries amounted to almost $40 billion compared with only $4 billion in 1977. Since 1981, concessionary savings have declined almost as rapidly; by 1985, they were only $8.5 billion.

Certain categories of developing countries maintained relatively large concessionary savings compared with others. In particular, debt-affected major borrowers and upper middle-income countries actually paid premiums for their credit by 1985. On the other hand, low-income countries, mostly in South Asia (figure 37) and Sub-Saharan Africa (figure 38), were still getting concessionary financing in 1985.

The loss of concessionary financing for the major debtor countries by 1985 is certainly one more factor that exacerbates the current debt problem.

THE CONSEQUENCES

The process of adjusting to the overaccumulation of debt in the 1970s has had several major consequences. Per capita income growth has declined, the direct result of policies to constrain imports, at least partially by inhibiting aggregate demand.[24] Trade also declined, a consequence of falling world and domestic income. Under a normal adjustment scenario where current account deficits are no longer sustainable, one would expect governments to undertake policies to constrain imports first and then undertake policies to stimulate exports. This reduction of imports was a major feature of the adjustment observed since 1982. However, exports have not grown as expected, partly because of reduced income growth in the developed countries. The resumption of renewed growth in the developing countries involves investment in new industries or investment in existing export industries to sustain export growth. The withdrawal of credit has been accompanied, and paid for, by reducing gross domestic investment.

The ability to generate renewed growth in developing countries is predicated on their capacity to increase exports. However, if substantial numbers of countries are simultaneously reducing capital formation as well as imports, increased export sales become extremely difficult, as has been the general case since 1982. Although many countries have been adjusting their current account balance, no evidence of renewed growth appears to be following it.

The adjustments to the debt crisis may well have forced developing countries (and, possibly, the world economy) into a low-level growth equilibrium. This situation will prevent the rapid reduction in the debt ratios which would lead to new credit availability and growth in the developing countries. Because these countries have been growth markets for U.S. agricultural exports, the main effect of the debt crisis has been to constrain world trade in general, agricultural trade as part of total trade, and U.S. agricultural exports as a major agricultural exporting nation.

Annual Changes in Real Per Capita Income

Real per capita income growth for the developing countries has declined since 1973. The debt-affected major borrowers have had particularly pronounced negative growth since 1981. The oil-exporting, middle-income countries (figure 39) had higher average growth in the early period but increasingly negative growth during 1982-85. The upper middle-income countries have had similar declining patterns over the period. The apparent increase in the per capita incomes of the low-income countries is almost completely the result of the large weight taken by India in that group, as is seen most clearly when contrasted with the performance of the Sub-Saharan countries of Africa, by far the most numerous countries of the low-income category (figure 40).

The Asian nations, in general, have had more positive growth patterns than other developing countries (figure 41). The Northeast Asian countries have had the highest real per capita income growth, compared with other groups, over the entire period. Although they have had an overall pattern of declining growth, their growth rates have increased in the first half of the 1980s, up to about 6 percent per year, very high by worldwide standards. Southeast Asia follows the more general pattern of declining growth, but to a modest degree. South Asia has actually had a pattern of increasing growth.

The African pattern was quite different (figure 40). Sub-Saharan Africa had increasingly negative per capita income growth. In 11 of the 12 years through 1985, these countries had absolute declines in real per capita GDP; the only year of positive growth was at less than 1 percent. The Sub-Saharan development problem is still the most challenging facing the world.

The North Africa and Middle East pattern followed closely that of the middle-income oil exporters with high early growth rates and high negative growth rates in the later part of the period. The change from an annual average growth of 10 percent to a negative growth of almost 5 percent was the greatest of any region.

Effect on Trade

From 1970 through 1980, both imports and exports of goods and nonfactor services increased rapidly. However, during 1975-83, the developing countries ran trade deficits. Between 1980 and 1981, imports rose as exports leveled off, generating a trade deficit for the developing countries of more than $80 billion. Between 1980 and 1984, imports declined by $76 billion, or a value almost equal to the 1980 trade deficit. Over the same period, exports increased $25 billion so that by 1984, there was a $20-billion trade surplus, a change of $100 billion from 1980. Between 1984 and 1985, both imports and exports fell so that the surplus declined to only $15 billion.

This pattern of declining imports and stagnant exports during 1980-84 is mirrored in almost all the trade patterns. The more critical the debt constraint the more dramatic the import curtailment and export promotion. Although imports declined by less than 20 percent for all developing countries, they declined by more than 30 percent for Latin American countries (figure 14) and 40 percent for debt-affected major borrowers. In Northeast Asia (figure 15), where the relative trade imbalance never became serious, both imports and exports increased, although exports increased more than imports. In Sub-Saharan Africa (figure 42), both imports and exports fell by 40 percent, a pattern also mirrored to some degree in North Africa and the Middle East (figure 43).

The Fall in Gross Domestic Capital Formation

One of the most pronounced features of 1970-85 was the increase and subsequent decrease in the rate of gross domestic capital formation. For all developing countries (figure 44), the rate averaged just over 23 percent during 1970-74, 27 percent during 1975-78, 26 percent during 1979-82, and then to 24 percent in 1984-85. The decline is most pronounced in the Latin American region and the countries comprising the debt-affected major borrowers. The fall in gross domestic capital formation is evident in all of the groupings except for those in Asia (figure 45), where the very high rates achieved in the middle of the period were exceeded by the end of the period.

The decline in gross domestic capital formation is one of the more pessimistic outcomes of the debt adjustment process. Without high rates of investment, renewed growth following the period of adjustment will be difficult.

Agricultural Trade

Agricultural trade patterns generally follow trends similar to those of total trade. The developing countries as a whole are net exporters of agricultural commodities. Between 1973 and 1981, imports rose faster (in nominal dollars) than exports. As with all exports, agricultural exports increased faster than imports during 1981-84, although the agricultural balance did not return to the levels of the late 1970s.

The peak year for imports was in 1981, with a decline following into 1984. Exports declined beginning in 1980, and only in 1984 returned to that level. Sub-Saharan Africa again had the bleakest picture, with both exports and imports significantly lower in 1984 than in 1980-81 (figure 46). South Asia's agricultural imports actually rose faster than exports. Southeast Asian imports of farm products remained steady (in dollar terms), with a sharp increase in exports in 1984 (figure 47). In that region, all the variation in the agricultural trade balance came from exports. The pattern was reversed in Northeast Asia, where exports remained constant, but imports declined (figure 48).

The pattern of Latin America, since 1982, was one of export promotion and import stagnation (figure 49). The dollar value of exports increased strongly after 1982, contributing a larger share to overall export earnings. Upper middle-income countries reversed the negative agricultural trade balance of 1980-81. The debt-affected major borrowers had the largest relative shift; imports remained at depressed levels after 1981, and exports increased most after 1982.

The general rule is that there have been no actual trend reversals, with the exception of Sub-Saharan Africa exports. The export trend increased for exporters, and imports stayed well above the levels of the late 1970s for all groupings.

Agricultural imports increased when compared with all imports by developing countries after 1982, rising to 15 percent of the total in 1984 from 13 percent in 1982. The most substantial increase was in Sub-Saharan Africa, where agricultural goods increased as a proportion of all imports since 1976. The immediate question is whether development goods are being sacrificed at ever-increasing rates as all imports decline.

The most dramatic case of agricultural imports supplanting other imports was in Latin America. Farm products rose to 15.5 percent of all imports, up from 11.5 percent in 1982, and higher than at any time during the 1970s. Only Northeast Asia sustained the trend of agricultural imports falling as a proportion of all imports. Major U.S. markets showed an upward trend in purchases of farm products in relation to all goods in 1982-84, up from 13 percent to 15 percent.

The other side of the picture is export promotion. Agricultural exports by all countries expanded about as fast as all exports (figure 50). Agricultural exports grew, especially in Sub-Saharan Africa, Southeast Asia (particularly Thailand and the Philippines), Latin America, and North Africa and the Middle East (figure 51). Those countries may be the ones in which expansion could be expected because they have traditionally been major producers of exportable crops. Major U.S. agricultural markets, on the other hand, continued to maintain agricultural exports as a constant proportion of all exports.

U.S. Agricultural Exports

U.S. exports to all 79 countries studied fell sharply in 1982, before recovering in 1983 and 1984 and plummeting again in 1985 (Table 9.2). The total dollar value in 1985 was only slightly above that of 1979. Only Sub-Saharan Africa imported a higher dollar value of agricultural products from the United States in 1985 than in 1984, for famine relief and under U.S. government programs.

The U.S. market share through 1984 remained above the levels of the late 1970s, except in 1982. Market share gains were confined to declining markets, however. U.S. farm products accounted for 50 percent of those in Latin America, up from 35-45 percent in the late 1970s. The United States maintained a larger proportion of total agricultural product sales in our major agricultural markets. The potentially expanding markets of the Asian regions have, however, been a loss in terms of U.S. agricultural penetration.

TABLE 9.2
Total Agricultural Imports and U.S. Share

Category	1970-75 Average	1976-80 Average	1981	1982	1983	1984	1985
World (million dollars)	92,725	196,080	253,900	234,300	228,800	239,400	230,437
U.S. Share (percent)	14.8	15.5	17.1	15.6	15.8	15.8	12.6
All 79 countries							
Total Agricultural Imports	15,585	35,725	56,064	48,747	48,815	51,988	47,112
U.S. Agricultural Share	27.8	26.7	27.4	25.4	28.7	27.7	24.6
Subsaharan Africa							
Total Agricultural Imports	1,606	4,006	6,174	5,517	4,912	4,829	4,752
U.S. Agricultural Share	10.6	13.5	16.5	14.8	13.2	15.3	21.9
South Asia							
Total Agricultural Imports	1,809	2,772	3,336	3,158	3,475	4,204	3,631
U.S. Agricultural Share	34.0	26.5	23.6	25.1	29.7	20.1	14.2
Southeast Asia							
Total Agricultural Imports	2,017	4,645	6,721	6,438	6,415	6,870	5,978
U.S. Agricultural Share	18.9	17.9	18.5	20.6	19.4	17.3	14.1
Northeast Asia							
Total Agricultural Imports	2,613	6,301	10,201	8,807	9,292	9,896	9,206
U.S. Agricultural Share	36.1	38.6	36.0	37.0	39.2	37.1	34.8
Latin America							
Total Agricultural Imports	3,214	7,590	12,463	9,509	9,170	9,269	8,129
U.S. Agricultural Share	43.6	42.5	49.7	44.9	54.7	54.3	49.3
North Africa/Middle East							
Total Agricultural Imports	3,642	9,160	15,690	14,008	14,465	15,765	14,360
U.S. Agricultural Share	19.4	17.4	14.8	12.4	14.9	17.4	12.9

Eastern Europe							
Total Agricultural Imports	683	1,253	1,478	1,310	1,086	1,155	1,057
U.S. Agricultural Share	13.3	11.4	9.2	13.8	24.7	16.3	12.7
Low Income							
Total Agricultural Imports	2,721	4,397	5,761	5,311	5,381	6,383	6,061
U.S. Agricultural Share	26.8	23.3	21.3	21.3	24.9	18.9	19.9
Middle Income							
Total Agricultural Imports	4,263	10,409	16,819	14,737	14,383	14,425	13,249
U.S. Agricultural Share	30.2	26.9	27.6	28.2	29.5	31.3	24.7
Upper-Middle Income							
Total Agricultural Imports	7,918	19,667	32,006	27,390	27,965	30,025	26,745
U.S. Agricultural Share	28.2	28.2	29.3	25.2	29.2	28.3	26.1
Oil Exporters							
Total Agricultural Imports	4,977	13,933	24,277	20,772	20,857	21,268	18,966
U.S. Agricultural Share	29.5	26.6	28.1	22.6	27.3	27.6	23.3
Major Borrowers							
Total Agricultural Imports	7,121	17,200	28,185	22,496	22,223	23,195	20,125
U.S. Agricultural Share	33.7	32.2	34.6	32.7	39.1	35.5	31.2
Debt-Affected Major Borrowers							
Total Agricultural Imports	3,283	7,894	12,710	9,679	8,971	8,664	7,610
U.S. Agricultural Share	29.1	29.8	36.7	32.3	42.2	45.4	37.5
Major U.S. Agricultural Markets							
Total Agricultural Imports	11,830	27,772	44,121	38,548	38,100	40,870	37,093
U.S. Agricultural Share	31.3	29.6	31.1	29.1	33.7	32.0	27.5

Source: Economic Research Service, U.S. Department of Agriculture.

THE CONSEQUENCES: AN ASSESSMENT

We can assess the probable consequences of the debt constraint by comparing actual outcomes of 1982-85 against simulations of outcomes over alternative financial constraint environments. (See Appendix B for the methodology).

Two very different patterns emerge for all developing countries. The groups which are the most debt constrained have actually generated outcomes which are below the full adjustment scenario, Latin America (figures 52 and 53), middle-income oil exporters, and Sub-Saharan African countries (figs. 54 and 55).

The Asian countries, on the other hand, appear to be only mildly constrained. Northeast and Southeast Asia are almost achieving GDP growth at an unconstrained result (figures 56 and 57), and South Asia is exceeding the projected constrained result. However, although these economies are achieving higher import growth in relation to potential outcomes, they also have had relatively slow import growth in line with the slower pace of growth in world trade (figures 58 and 59).

A comparison of the unconstrained estimates and the historical values between 1983 and 1985 gives an indication of the short-run costs of the crisis (Table 9.3). All 79 countries show a cumulative loss of $392 billion in real GNP between 1983 and 1985. Imports to those same countries were reduced by an aggregate $379 billion.

A particularly interesting feature of Table 9.3 is the potential interpretation of the adjustment to the crisis by the two most disparate sets of countries: the debt-affected major borrowers as opposed to Northeast Asia. The new balance-of-payments equilibrium achieved by the debt-affected group, through import constrained and with negative net transfers, has been heavily reliant on declines in domestic income. The Northeast Asian outcome, on the other hand, may be the result of a far different phenomenon. Much of the fall in credit availability could have been accommodated through the higher domestic savings rates of these countries. Instead of declines in domestic income, the adjustment would require a larger reduction in consumption of importables. GDP is barely affected, but imports have declined almost to the level of the full adjustment scenario (figures 56 and 58). The notion that the debt crisis has not affected the Northeast Asian countries is clearly insupportable. In spite of not being constrained by repayment problems, they took preventive measures before any problem appeared. This clearly implies that they are exercising better macro planning than the debt affected countries.

TABLE 9.3
Historical Costs of Financial Constraints, Measured as the Difference Between Unconstrained and Historical Outcomes for Gross Domestic Product and Imports, in Billions of 1982 Dollars, 1983-85

Category	GDP	Imports
All Developing Countries	392.3	378.8
Major Borrowers	310.7	191.2
Debt Affected Major Borrowers	252.1	121.2
Non Debt Affected	58.6	70.0
North Africa & Middle East	65.2	94.8
Subsaharan Africa	60.7	46.9
Northeast Asia	11.7	29.3
South Asia	-16.1	9.4
Southeast Asia	10.7	57.5
Latin America	246.8	132.6

Source: Economic Research Service, U. S. Department of Agriculture.

THE RESOLUTION OF THE DEBT PROBLEM

The ideal world scenario for resolving the debt crisis would include a period in which debt-affected countries would undertake policy changes to realign their export-import balance followed by a period of renewed world growth led by expansion of trade. However, there is no evidence of this actually occurring. The realignment that occurred in the most debt-affected countries occurred at far greater than zero cost. Further, world growth (with the exception of the United States) has been nowhere close to rates achieved in the 1970s.

Except for North Africa and the Middle East and South Asia, the needed adjustment to the change in finance availability has taken place, but there is scant evidence that this adjustment will be followed by renewed income and trade growth. The global effect of contracted imports and export promotion in such a large part of the world has led to a situation in which the export markets have become more competitive and more constrained.

Commodity prices, among other goods traded internationally, have fallen. The United States is now undertaking policies to reduce its high trade deficit that was present in most of the study period. Japan is running an $80-billion trade surplus and could provide a growing export market, but it seems unwilling to take the required steps.

Financial institutions are also unwilling or unable, on net, to further lend to the developing countries. Commercial lenders have been withdrawing credit from these countries through the process of negative net transfers in excess of $30 billion by 1985.

Solutions to date have served to maintain the present value of developing country debt. Rescheduling debt has become commonplace with the effect of superficially improving the term structure of the debt but not of reducing its burden. The debtor countries find themselves in a situation where the debt load is equal to or greater than it was at the start of the debt crisis in 1982. For all of the adjustments and renegotiations, the constraint which debt has imposed on world trade and development has not been noticeably reduced.

Recently solutions which truly reduce the burden of debt have, in fact, begun to be implemented. The possibility of debt for equity swaps, and the development of secondary markets for Third World debt are elements of this solution. The recent development with Mexico where they will swap outstanding loans for bonds backed by U.S. "zero coupon" bonds is further progress in this direction. The fact that financial institutions are now more willing to consider writedowns and writeoffs of parts of developing countries debts is further made credible by the fact that the exposure of the commercial banks is now so much less than it was in 1982. All of these recent tendencies hold out promise that significant steps towards reducing the financial constraints to trade may at last be forthcoming. However, much serious negotiations are still ahead in terms of how the cost of the writedowns will be shared between the developing countries, the lending institutions and the developed countries. The success of these negotiations will determine whether the seriously debt impacted countries will get a new opportunity to initiate sustained development and trade growth.

NOTES

1. The technical term for the difference between new borrowing and total repayments is "net transfer." However, "net credit flow" seems a more apt description.

2. Prices for all raw materials, not only oil, generally increased during 1973-81. Although other resources did not, in general, increase in value as drastically as did oil, their increases were nonetheless dramatic. Examples, for 1972-80, include a quadrupling of the dollar prices of bauxite and rubber, a tripling of prices for aluminum and coffee, and a doubling of prices for nickel, copper, and manganese (Shane, 1986).

3. The average growth rate of U.S. M1 was 6.7 percent during 1971-81, but 8.2 percent over the next 5 years.

4. The terms world liquidity, Eurocurrencies, and overseas assets will be used interchangeably throughout.

5. The growth rates of U.S. M1, industrial country M1, and all countries' M1 (figure 4) bear little relationship to the growth rate in overseas bank assets, most notably between 1981 and 1985. If one looked at U.S., industrial countries', and all countries' M1 between 1981 and 1985 the only conclusion would have been that money was more readily available during that time, and that the 1970s were noted for restraint.

6. Several reasons have been suggested, but the most plausible explanation is that increased wealth from Government bond issues was followed by a 3-year growth market in stocks. This situation would also be complemented by increased retained earnings by the corporate sector in response to higher real borrowing rates during the period. The shift is reflected in the downward movement in income velocity between 1980 and 1983 estimated as M2 (M1 + savings deposits without checking privileges + time deposits of less than $100,000) divided by gross national product (GNP) (International Monetary Fund, 1987). The 3-year decline is unprecedented for the post-World War II period.

7. "Sterilization" is a process by which the central monetary authority (the Federal Reserve in the United States) takes action to counter otherwise automatic changes in the domestic money stock as a result of efforts to maintain a fixed exchange rate. A currency outflow that would result from a balance-of-payments deficit would, in the United States, be offset by an open market purchase of Government securities, leaving the domestic money stock unchanged.

8. This situation could conceivably occur even when the rate of growth in domestic money exceeded that of reserve accumulation. The real appreciation would still be the result of a domestic rate of return that is higher than the "world" rate.

9. One could maintain, however, that reserves were increased to improve creditworthiness.

10. The strength of the dollar has reversed, of course, since 1985, but mostly against the currencies of OECD countries. The dollar has actually appreciated against the debtor countries since 1985 into early 1988.

11. For the middle-income countries, a logarithmic scale with Bolivia and a cardinal scale without Bolivia produce very similar trends.

12. Calculated by the International Monetary Fund. See International Financial Statistics for historical data.

13. The four countries during 1956-65 were Argentina, Turkey, Brazil, and Chile. The seven countries during 1966-75 were Cambodia, Chile, Ghana, India, Indonesia, Pakistan, and Peru.

14. The 11 countries during 1976-80 were Bolivia, Jamaica, India, Liberia, Nicaragua, Peru, Sierra Leone, Sudan, Togo, Turkey, and Zaire.

15. The 25 countries rescheduling during 1981-83 were Argentina, Bolivia, Brazil, Central African Republic, Chile, Costa Rica, Cuba, Ecuador, Guyana, Jamaica, Liberia, Madagascar, Malawi, Mexico, Nicaragua, Pakistan, Poland, Romania, Senegal, Sudan, Togo, Turkey, Uganda, Yugoslavia, and Zaire.

16. The 18 countries rescheduling in 1984 were Brazil, Ecuador, Ivory Coast, Jamaica, Liberia, Madagascar, Mozambique, Nicaragua, Niger, Nigeria, Peru, Philippines, Senegal, Sierra Leone, Sudan, Togo, Yugoslavia, and Zambia.

17. The 24 countries rescheduling in 1985 were Argentina, Bolivia, Central African Republic, Chile, Costa Rica, Dominican Republic, Ecuador, Equatorial Guinea, Ivory Coast, Jamaica, Madagascar, Mauritania, Mexico, Morocco, Niger, Panama, Peru, Philippines, Senegal, Somalia, Sudan, Togo, Yugoslavia, and Zaire.

18. Net transfers is defined as disbursements less total debt service and is equal to the change in total debt less interest payments.

19. Overall short-term balance-of-payments equilibrium requires that the capital account equal the negative of the current account balance. A reduction in capital inflow (net transfers) must be accompanied by a fall in net imports.

20. The net adjustment (NA) is NA = X - M - iD, where X = exports of goods and nonfactor services, M = imports of goods and nonfactor services, i = the current interest rate on the level of total debt, D. The adjustment rate is then NA/X. All magnitudes are nominal.

21. This situation is true except to the degree that the real interest rate is lowered through renegotiation. There is no evidence, however, through the study period, that interest rates have been reduced. Indeed, the typical rescheduling raised the spread over LIBOR.

22. The increases were not as spectacular as during 1979-82. However, there was no evident tendency for the debt/export or debt/GDP ratios to decline. The burden was not lifted.

23. There is often a spread between the long- and short-term rates. Over time, the two rates, if of equal risk, should be equal.

24. Many countries responded to their balance-of-payments deficits by implicitly acknowledging the possibility that excess aggregate demand (in the form of fiscal deficits or excessive inflation) contributed to increased imports. The policies implemented to reduce aggregate demand included fiscal and monetary restraint combined with exchange rate depreciations or other trade policy measures. The consequently reduced import demand was therefore accompanied by declining income.

REFERENCES

Abbott, P. C. Foreign Exchange Constraints to Trade and Development. FAER-209. Economic Research Service, U. S. Department of Agriculture. Washington, D. C.

Executive Office of the President. Economic Report of the President. Washington, D.C.: U. S. Government Printing Office, 1986.

Feldstein, Martin. International Debt Policy. Council of Economic Advisors, May 1984.

Food and Agriculture Organization of the United Nations. Trade Yearbook Tape. Rome, 1986.

Gerrard, C. and T. Roe. "Government Intervention in Food Grain Markets," Journal of Development Economics. 12(1983): 109-32.

Hayami, Y. and V. W. Ruttan. Agricultural Development: An International Perspective. Baltimore: Johns Hopkins University Press, 1985.

International Monetary Fund. International Financial Statistics Tape. Washington, DC, July 1987.

_______. External Debt in Perspective. Washington, DC, 1983.

_______. Interest Rate Policies in Developing Countries. Occasional Paper 22. Washington, D.C. October 1983.

_______. Recent Multilateral Debt Restructuring with Official and Bank Creditors. Occasional Paper 25. Washington, D.C., December 1983.

_______. World Economic Outlook, 1984. Washington, DC, 1984.

Jabara, C. Terms of Trade for Developing Countries. FAER-161. Economic Research Service, U. S. Department of Agriculture. Washington, D. C. November 1980.

Khan, M.S., and M. Knight. "Sources of Payments Problems in LDCs," Finance and Development. 20(December 1983): 2-5.

Killick, T. and M. Sutton. "Disequilibria, Financing, and Adjustment in Developing Countries," Adjustment and Financing in the Developing World. Tony Killick, (ed.) Washington, D.C.: International Monetary Fund in association with the Overseas Development Institute (London), 1982.

Lee, J. E., Jr., and M. D. Shane. "U.S. Agricultural Interests and Growth in the Developing Economies: The Critical Linkage." U. S. Agriculture and Third World Economic Development: Critical Interdependency, National Planning Association, Washington,D. C. February 1987, pp. 52-78.

Mundell, R. Monetary Theory. Pacific Palisades, CA: Goodyear Publishing, 1971.

Organization for Economic Cooperation and Development. External Debt of Developing Countries. Paris, 1982.

Roe, T., M. D. Shane, and De Huu Vo. Price Responsiveness of World Grain Markets: The Influence of Government Intervention on Import Price Elasticity. TB-1720. Economic Research Service, U. S. Department of Agriculture. Washington, D. C.

Shane, M. "Capital Markets and the Dynamic of Growth," American Economic Review. 64(March 1973):162-69.

_______. "Government Intervention, Financial Constraint and the World Food Situation." Paper presented at the Conference on Food Policies and Politics, Purdue Univ., West Lafayette, IN, May 12, 1986.

_______, and D. Stallings. Financial Constraints to Trade and Growth: The World Debt Crisis and Its Aftermath. FAER-211. Economic Research Service, U. S. Department of Agriculture. Washington, D. C. December 1984a.

_______, and D. Stallings. Trade and Growth of Developing Countries Under Financial Constraint. Staff Report AGES840519. Economic Research Service, U.S. Department of Agriculture. Washington, D. C. June 1984b.

United Nations. World Economic Survey, 1986. New York, June 1986.

World Bank. World Debt Tables, 1985-86. Washington, DC, March 1986.

_______. World Development Report 1985. New York: Oxford Univ. Press, 1985.

_______. World Tables Tape Revised. Washington, DC, 1985.

APPENDIX A
Country Categories

Region/country	Low income economies	Middle-income economies	Upper middle-income economies	Oil Exporters	Major borrowers 1/	Debt-affected major borrowers 2/	Major U.S. agricultural markets 3/
North Africa/Middle East							
Algeria			1	1			1
Egypt		1		1	1		1
Iran			1	1			
Iraq			1	1			1
Jordan			1				
Lebanon		1					
Morocco		1		1	1	1	1
Syria			1				
Tunisia		1					1
Turkey		1		1	1		1
Yemen Arab Rep. (Sana)		1					
Subsaharan Africa							
Benin	1						
Burkina Faso	1						
Cameroon		1		1			
Central African Rep.	1						
Chad	1						
Ethiopia	1						
Gabon		1		1			
Gambia	1						
Ghana	1						
Ivory Coast		1					
Kenya	1						
Liberia	1						
Madagascar	1						
Malawi	1						

Mali	1						
Mauritania		1					
Mauritius	1						
Niger	1						
Nigeria		1		1	1	1	1
Rwanda	1						
Senegal		1					
Sierra Leone	1						
Somalia	1						
Sudan	1						
Tanzania	1						
Togo	1						
Uganda	1						
Zaire	1						
Zambia		1					
Zimbabwe		1					
Northeast Asia							
Hong Kong			1				1
Korea, Rep. of			1		1		1
Taiwan			1				1
South Asia							
Bangladesh	1						1
Burma	1						
India	1				1		1
Nepal	1						
Pakistan	1						1
Sri Lanka	1						
Southeast Asia							
Indonesia		1		1	1		1
Malaysia			1	1	1		1
Papua New Guinea		1					
Philippines		1		1	1	1	1
Singapore			1	1			1
Thailand		1			1		1

(continued)

APPENDIX A

Country Categories (Cont.)

Region/country	Low income economies	Middle-income economies	Upper middle-income economies	Oil Exporters	Major borrowers 1/	Debt-affected major borrowers 2/	Major U.S. agricultural markets 3/
Latin America							
Argentina			1	1	1	1	
Bahamas			1				
Bolivia		1					
Brazil			1		1	1	1
Chile			1		1	1	1
Colombia		1		1			1
Costa Rica		1					
Dominican Republic		1		1			1
Ecuador		1		1			1
El Salvador		1		1			1
Guatemala		1					
Guyana	1						
Haiti	1			1			1
Honduras		1					
Jamaica		1		1			1
Mexico			1	1	1	1	1
Nicaragua		1					
Panama			1	1			1
Paraguay		1		1			1
Peru		1		1			1
Uruguay		1					
Venezuela			1	1	1		1
Yugoslavia			1		1		1

1/ Over $10 billion in all external debt.
2/ Rescheduled during 1982-86.
3/ Purchases of at least $200 million of U.S. farm products in any 3-year period.

APPENDIX B: METHODOLOGY

The macrofinancial simulations are derived by introducing varying degrees of financial constraint into the simple open economy macroeconomic general equilibrium model described below.

We established the macrofinancial simulations model with the following values, derived from existing data at time t_0 = 1982:

(1) $Y_0 = \underline{Y}$

(2) $I_0 = \underline{I}$

(3) $X_0 = \underline{X}$

(4) $M_0 = \underline{M}$

(5) $D_0 = \underline{D}$

where the variables are defined as

Y = GNP in U.S. dollars

I = Gross domestic capital formation in U.S. dollars

X = Exports of nonfactor goods and services

M = Imports of nonfactor goods and services

D = Total disbursed external debt (public and private)

Using the above initial values, the growth rate of GNP for year t_1 is derived in the following way:

(6) $\dot{Y} = pt\dot{K}$

where the (.) refers to the time derivations of the variable interpreted for empirical purposes as the annual change, and

p = the marginal product of capital

t = embodied rate of technical change

Given a fixed depreciation rate for capital, d,

(7) $\dot{K} = (I - dK)$

(6') $\dot{Y} = pt(I - dK)$.

Because country-specific measures of the capital stock (K) are not readily available, the depreciation rate is taken as proportionate to the rate of capital formation.

(8) $d = a(I/K)$

Furthermore, to allow country-specific variation in the depreciation rate depending upon the rate of (annual) capital formation, the proportionality factor (a) is taken to be a linear function of the investment rate:

(9) $a = c_0 + c(I/K)$

This gives a modified growth equation:

(6") $\dot{Y} = (pt - a)I$

(10) $\dot{y} = (pt - a)(I/Y)$

where $\dot{y} = \dot{Y}/Y$. (I/Y) has an initial lagged value of historical data.

Unconstrained simulations of imports (M) and exports (X) are derived by multiplying elasticity estimates (m and c, respectively) by growth estimates as follows:

(11) $\dot{m} = m\dot{y}$

(12) $\dot{x} = \dot{y}_1$

where $\dot{y}_1$ is an assumed growth rate of GNP in the industrial countries.

Financial constraints enter the model in two ways: By reducing investment and by reducing the growth rate of income. An adjustment requirement or unpaid residual is calculated by computing the amount by which net exports differ from required interest payments. This residual is positive if interest payments cannot be made entirely out of a net trade surplus and zero otherwise. Thus,

(13) $A(t) = M(t) - X(t) + rD(t\text{-}1)$ if "unpaid residual" > 0
$= 0$ otherwise

($rD(t\text{-}1)$ refers to interest owed on external debt outstanding).

In the base scenario, in which no external financing is forthcoming, we assumed that adjustments in capital formation, trade, and growth must be made so that A(t) goes to zero in the first year of the simulation. In the partial adjustment cases, we assumed that only a fraction of this adjustment needs to be undertaken in any year. These changes are incorporated into the model through equations (10), (11), and (12). Thus,

(14) $(I^*/Y) = (1 - aU)(I/Y)$

where (I^*/Y) is the adjusted rate of investment, $a = A(t)/Y(t)$, and U is the proportion of the "unpaid residual" which is deferred. Changes in (I^*/Y) will modify y and thus m.

By varying the adjustment rate U between zero and one, we simulate an alternative financial constraint scenario. The case of U = 0 (the "full adjustment" scenario) corresponds to the "trade-constrained" case described in (Abbott). The case of U = 1 corresponds to the "savings-constrained" case.

Figure 1
U.S. domestic and overseas interest rates

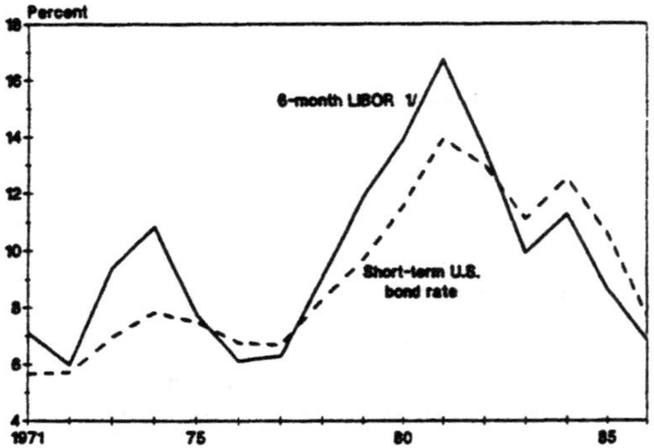

1/ London Interbank Offer Rate

Figure 2
World M1 vs. overseas bank assets

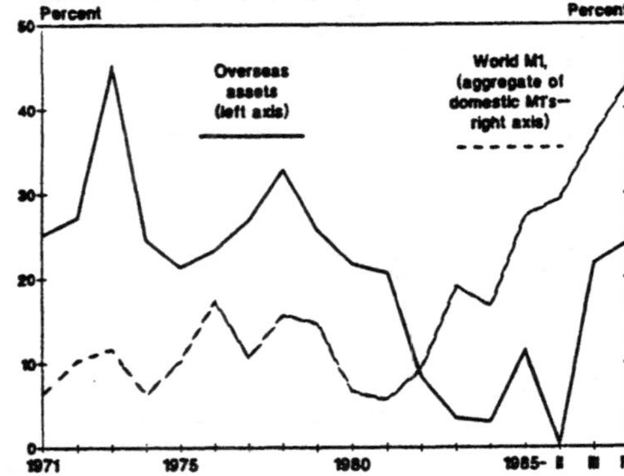

Figure 3
Growth in debt matches growth in world money

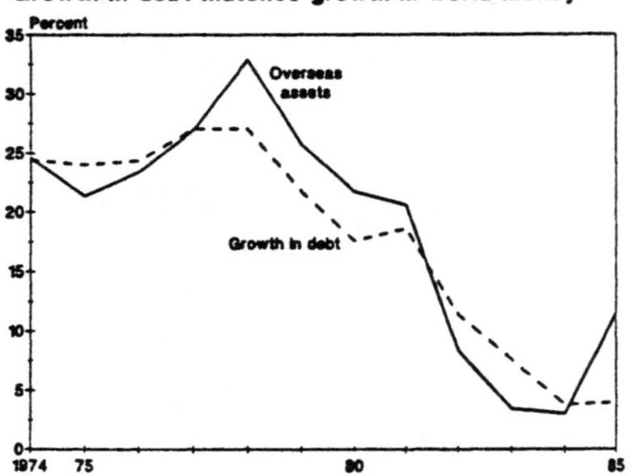

Figure 4
Reserves decline as money growth reduced

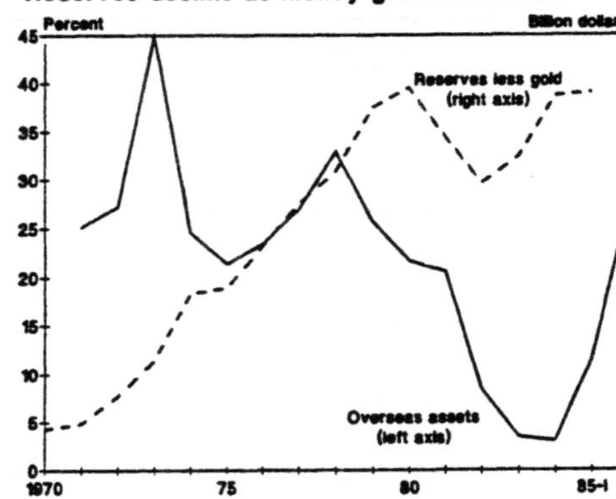

Figure 5

Reserves in selected reserve - loss countries

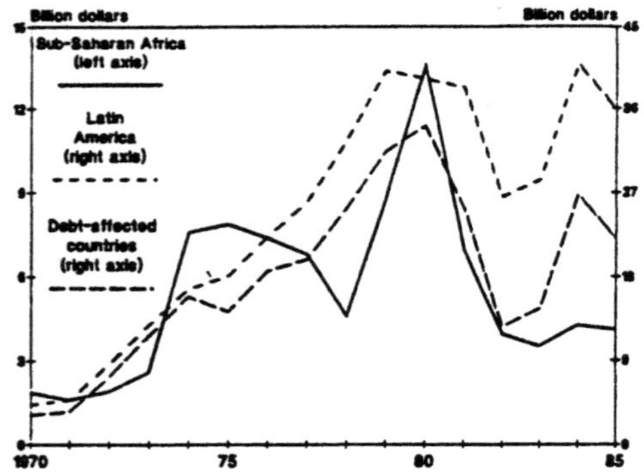

Figure 6

Reserves and current account balances: All 79 countries

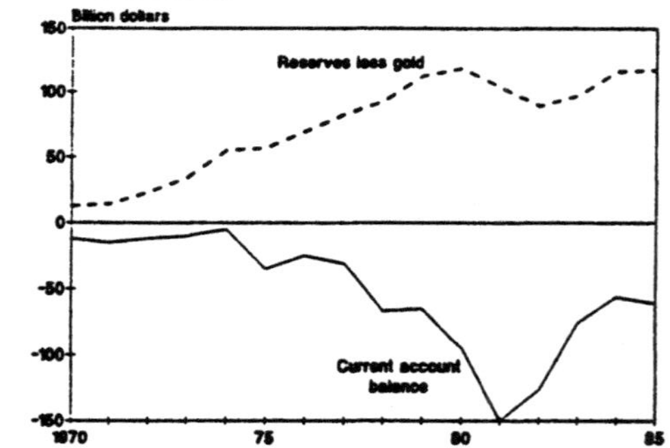

Figure 7

Reserve/import and reserve/export ratios: All 79 countries

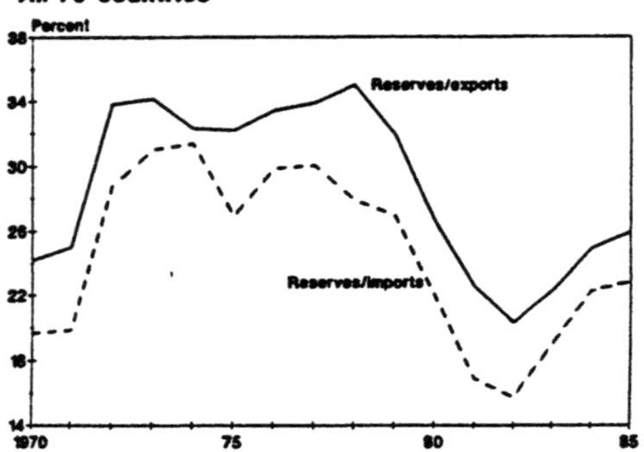

Figure 8

Reserve/import ratios: Northeast Asia and Latin America

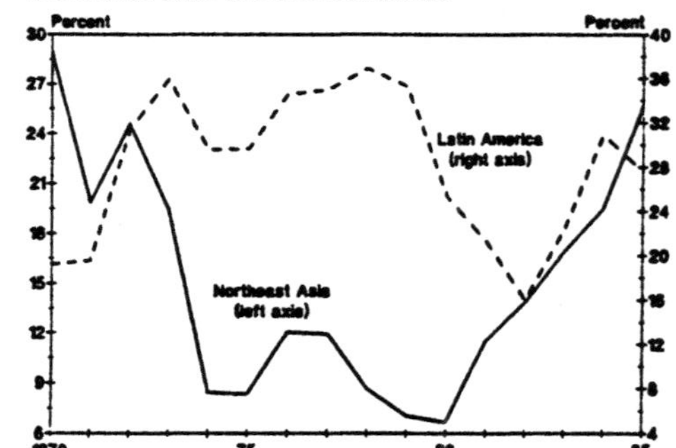

Figure 9
Real interest rate, all 79 countries:
LIBOR 6-month less change in export unit values 1/

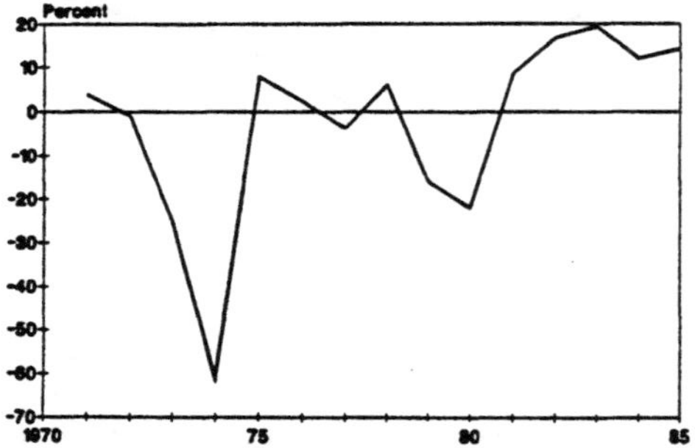

1/ LIBOR = London Interbank Offer Rate

Figure 10.
Real exchange rate, all 79 countries, Latin America, and major U.S. markets, deflated by CPI's

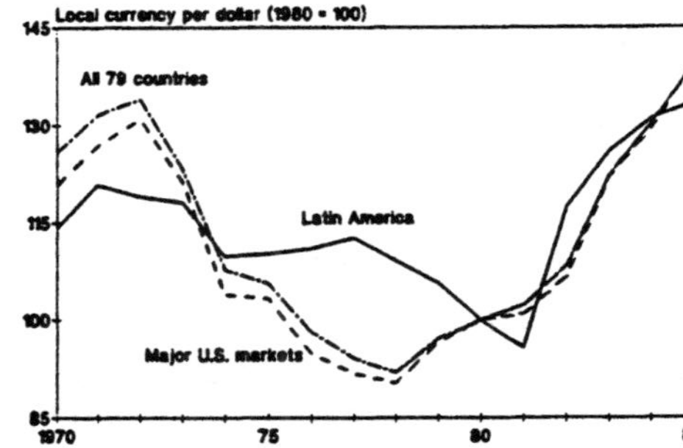

Figure 11
Consumer price index, all 79 countries, with and without Bolivia

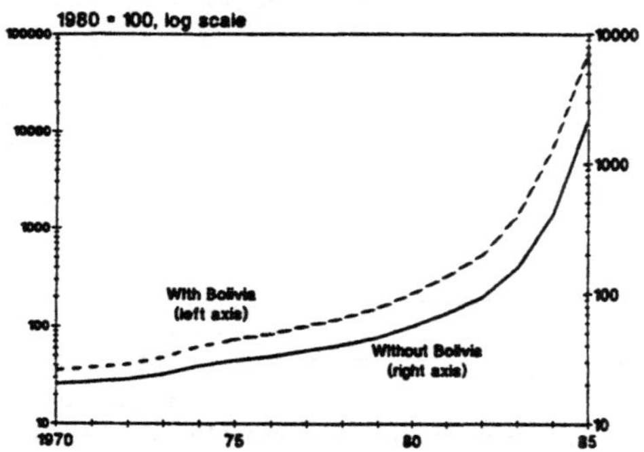

Figure 12
Change in consumer price index and gross capital formation as a share of gross domestic product, debt-affected countries

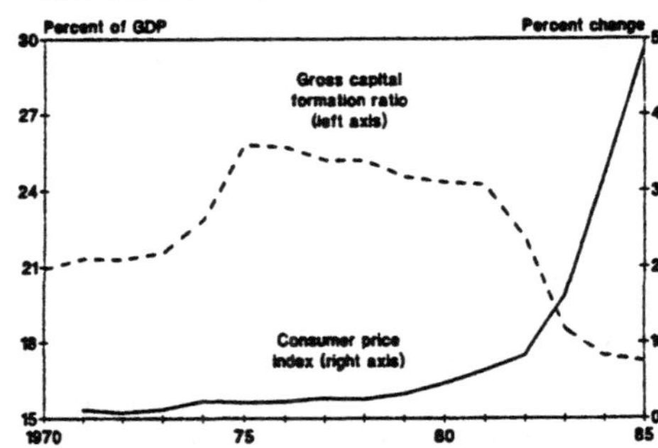

Figure 13
Change in export prices and real exchange rates, all 79 countries

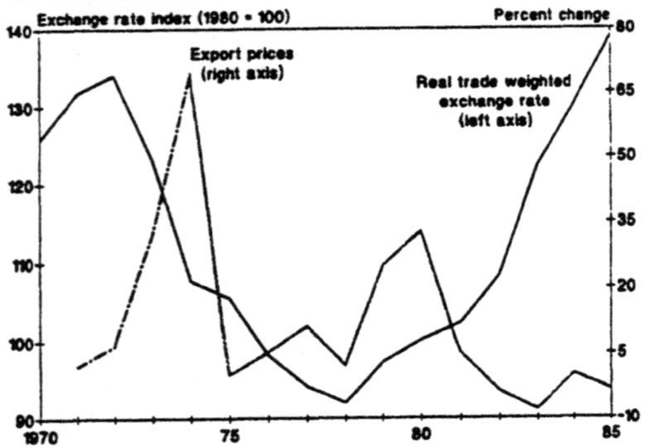

Figure 14
Trade in goods and nonfactor services, Latin America

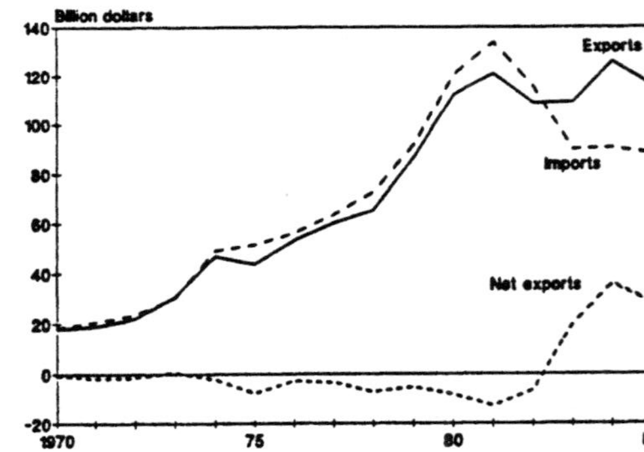

Figure 15
Trade in goods and nonfactor services, Northeast Asia

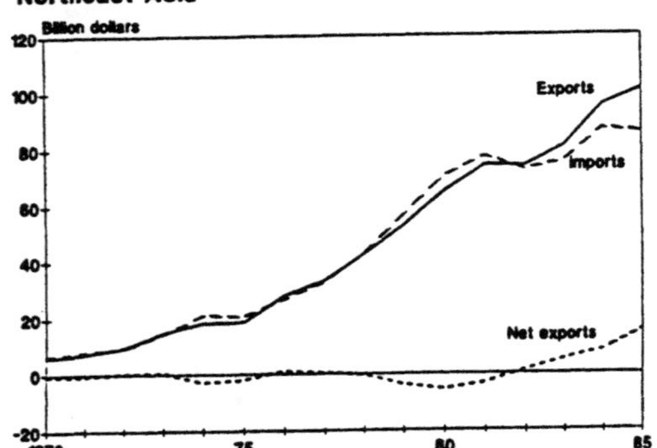

Figure 16
Real merchandise imports, all 79 countries and Latin America

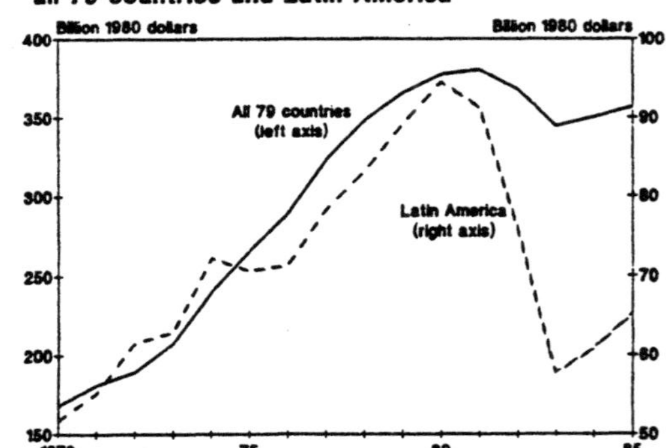

Figure 17
Real merchandise imports, Northeast Asia and Southeast Asia

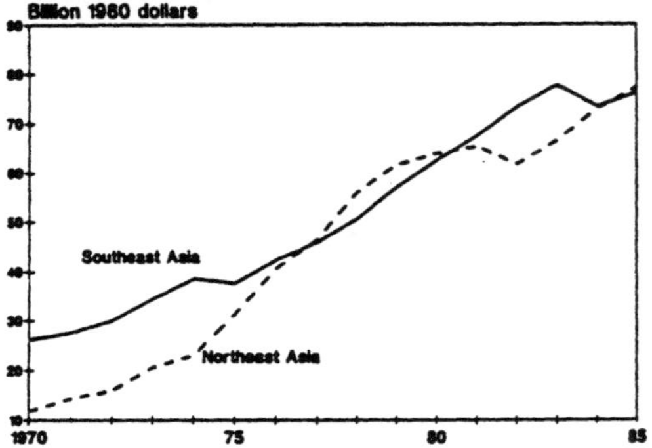

Figure 18
Declining real merchandise exports, Sub-Saharan Africa and oil exporters

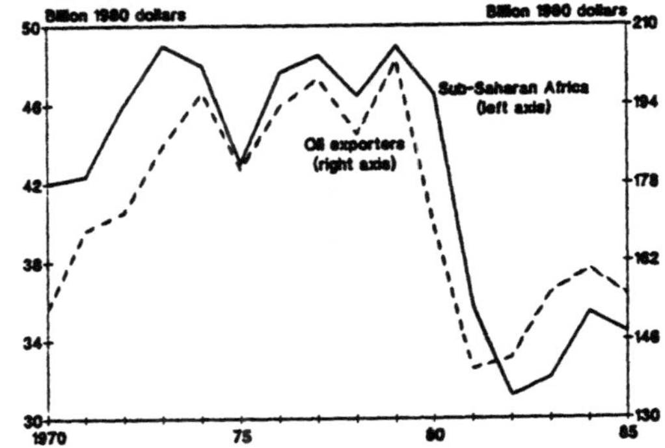

Figure 19
Increased real merchandise exports, South Asia, Southeast Asia, and Northeast Asia

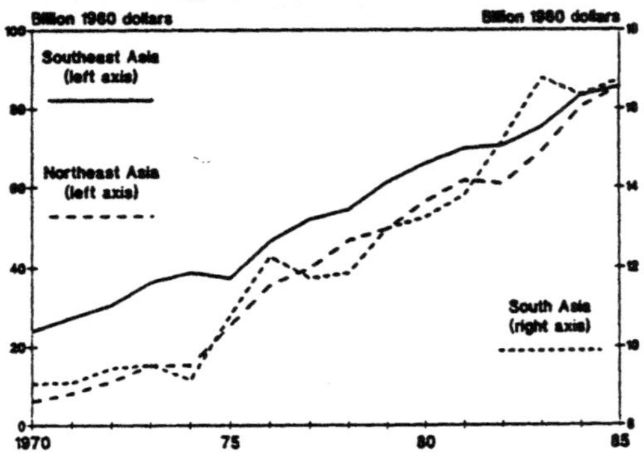

Figure 20
Real services deficit, Latin America and debt-affected major borrowers

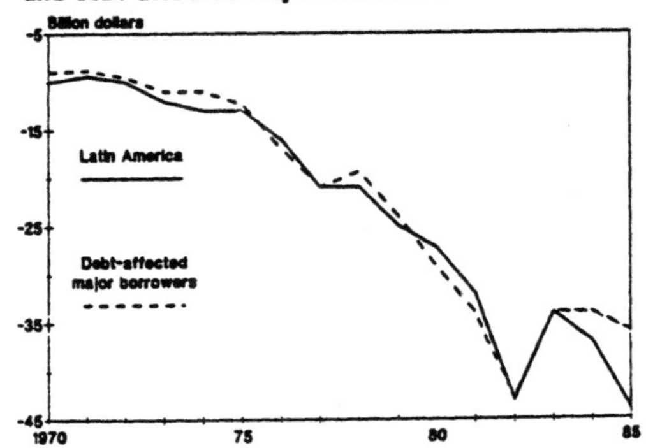

Figure 21
Debt reschedulings, number of countries rescheduling, and type of debt

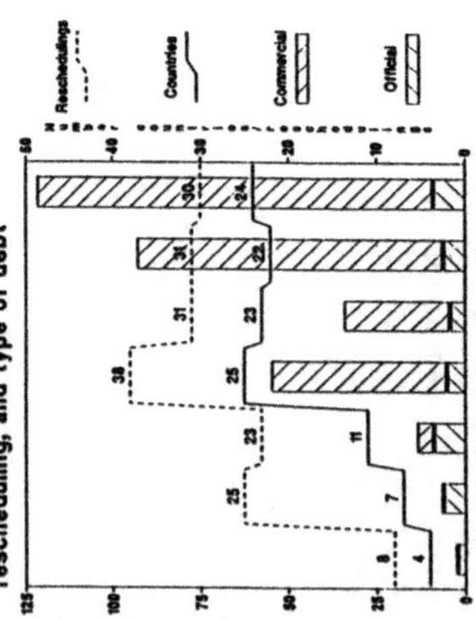

Figure 22
Structure of total debt, Northeast Asia

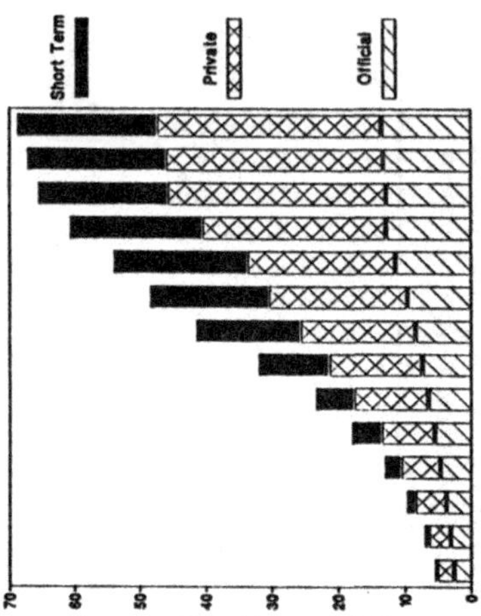

Figure 23
Structure of total debt, low income economies

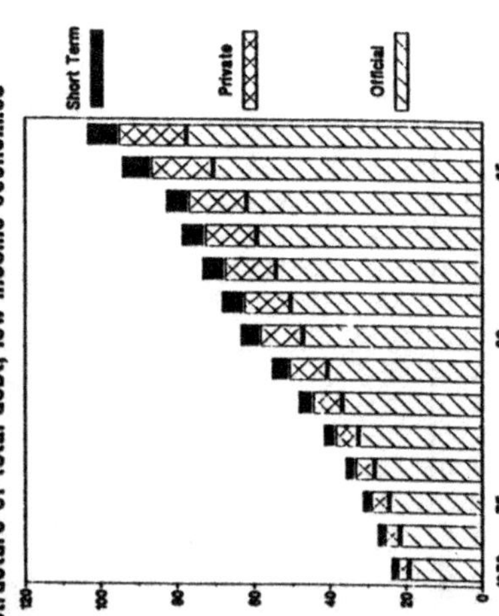

Figure 24
Structure of total debt, Latin America

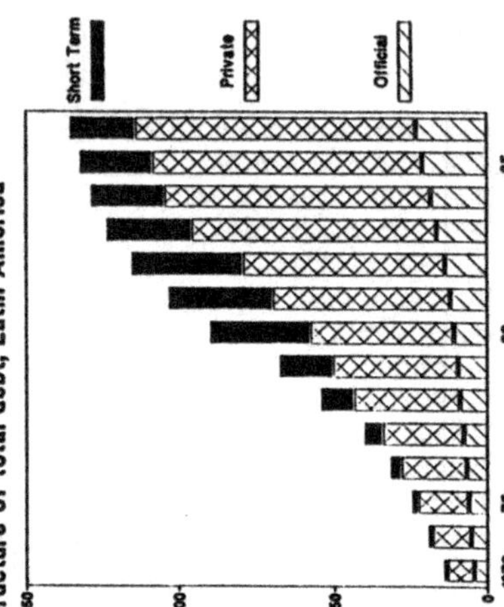

Figure 25
Distribution of total debt, geographic regions

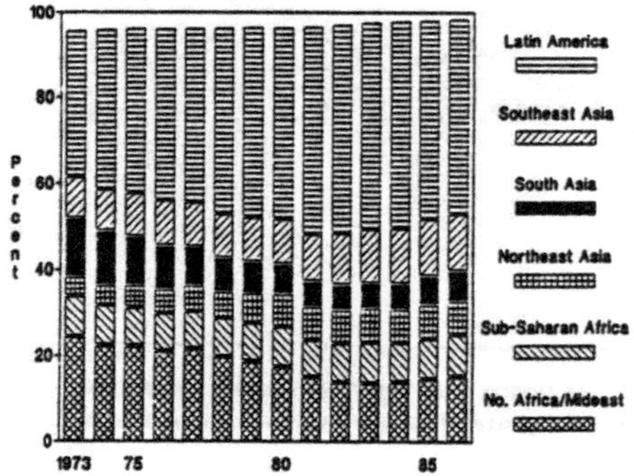

Figure 26
Distribution of total debt, income groups

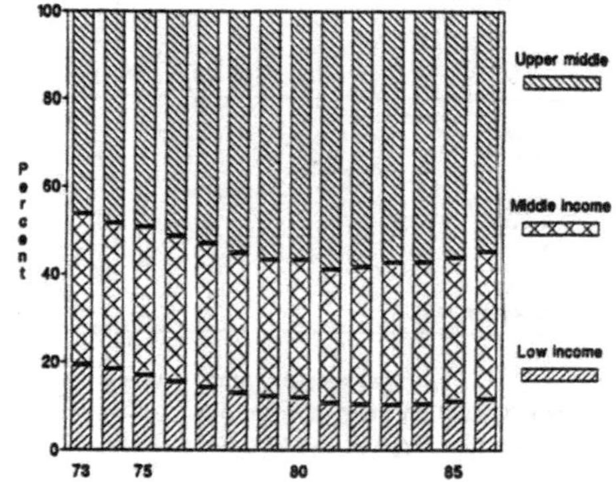

Figure 27
Annual change in total debt, all 79 countries

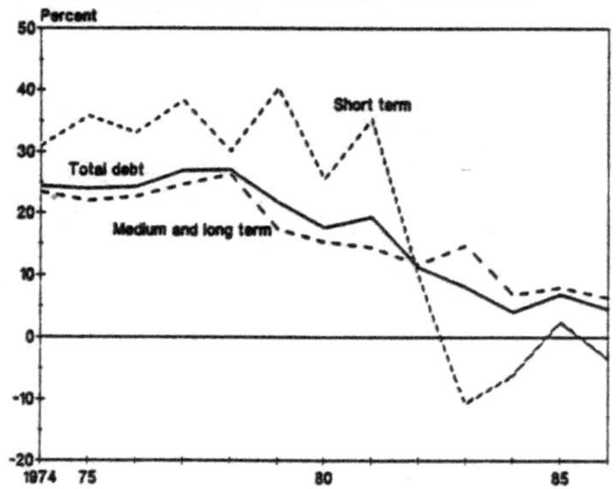

Figure 28
Net transfers, all 79 countries, U.S. agricultural markets, and oil exporters 1/

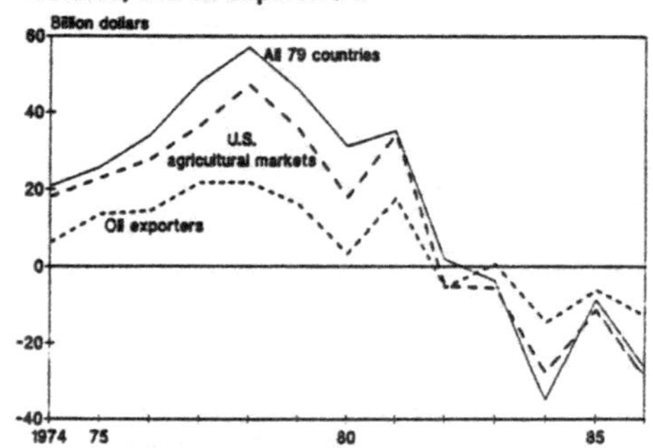

1/ New borrowing less debt service payments.

Figure 29

Net transfers, Latin America, Sub-Saharan Africa, and North Africa/Middle East 1/

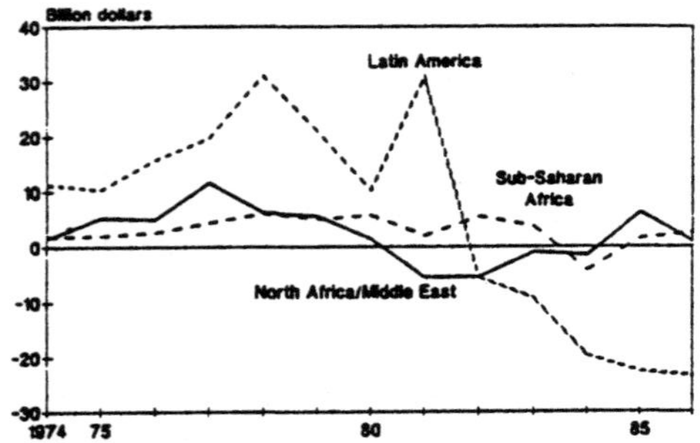

1/ New borrowing less debt service payments.

Figure 30

Net exports needed to offset interest payments, all 79 countries, non-debt and debt-affected countries

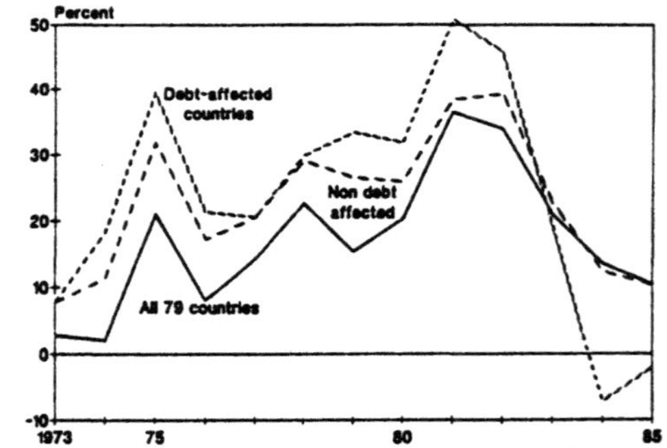

Figure 31

Net exports needed to offset interest payments, South Asia, Southeast Asia, and Northeast Asia

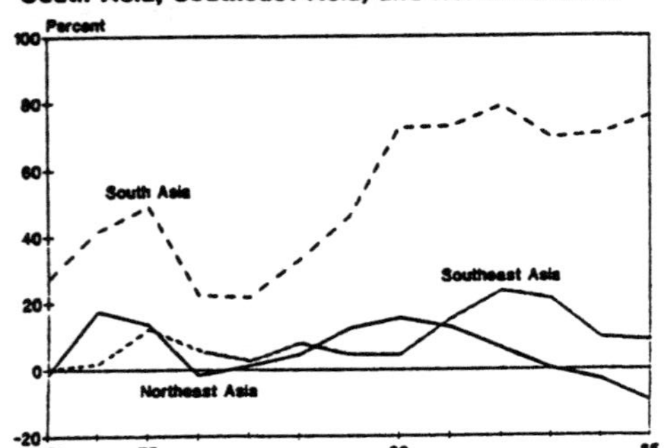

Figure 32

Net exports needed to offset interest payments, Latin America, Sub-Saharan Africa, and North Africa/Middle East

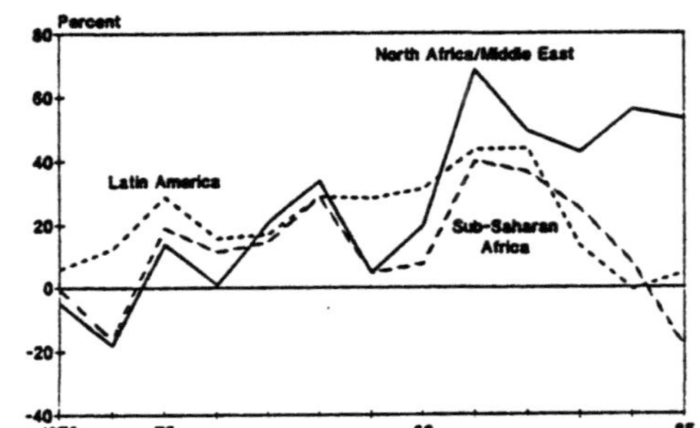

Figure 33

Debt service as a share of total exports, Latin America, Sub-Saharan Africa, and North Africa/Middle East

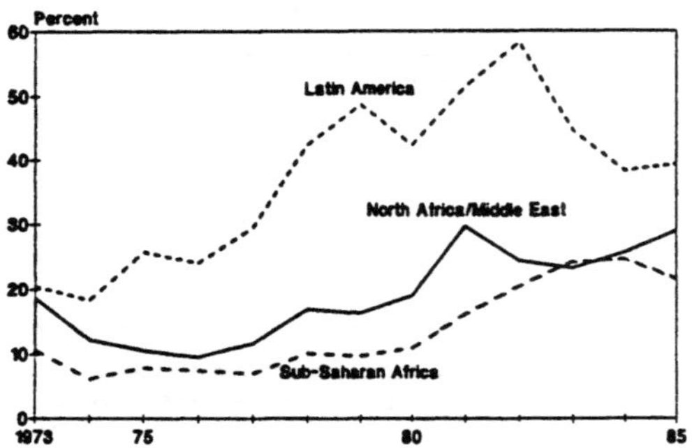

Figure 34

Debt service as a share of total exports, South Asia, Southeast Asia, and Northeast Asia

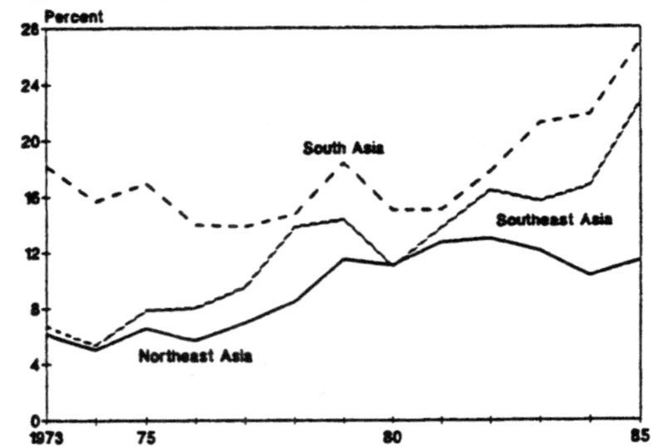

Figure 35

Total debt as a share of gross domestic product, all 79 countries, U.S. agricultural markets, and oil exporters

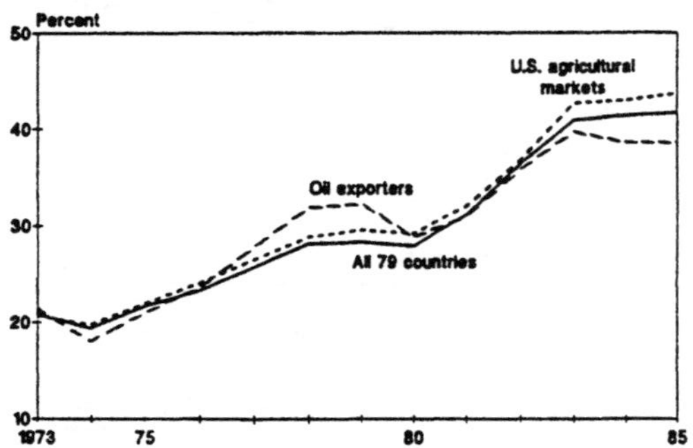

Figure 36

Total debt as a share of total exports, all 79 countries, U.S. agricultural markets, and oil exporter

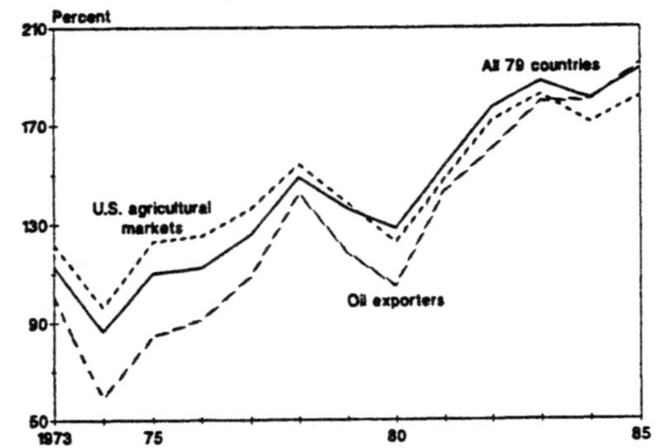

Figure 37

Savings from concessionary interest rates on medium- and long-term debt, South Asia, Southeast Asia, and Northeast Asia

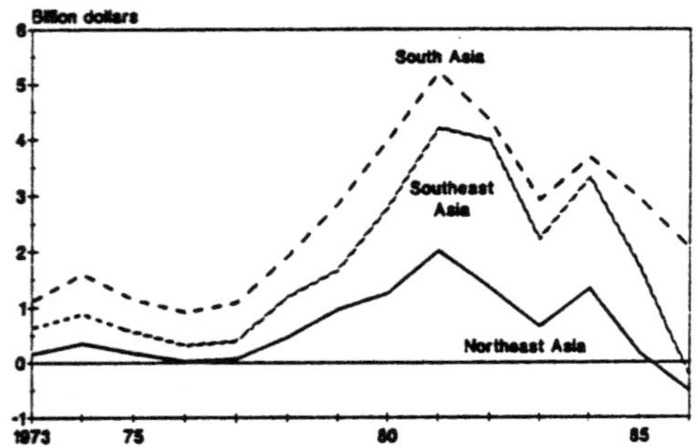

Figure 38

Savings from concessionary interest rates on medium- and long-term debt, Latin America, Sub-Saharan Africa, and North Africa/Middle East

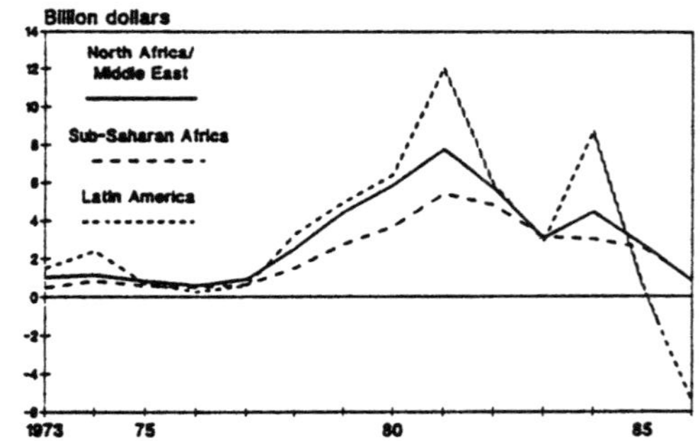

Figure 39

Annual change in real per capita income, all 79 countries, U.S. agricultural markets, and oil exporters

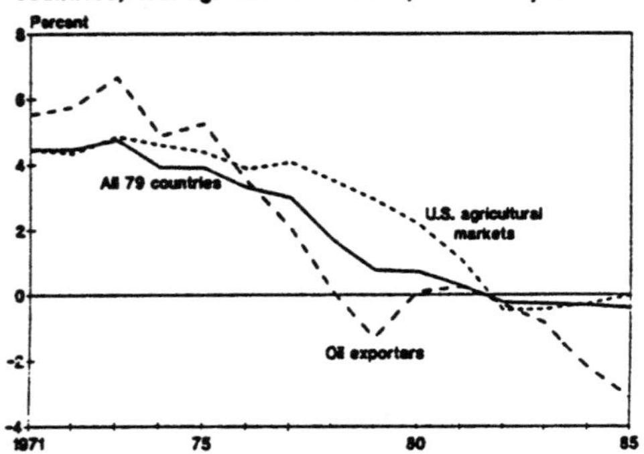

Figure 40

Annual change in real per capita income, Latin America, Sub-Saharan Africa, and North Africa/Middle East

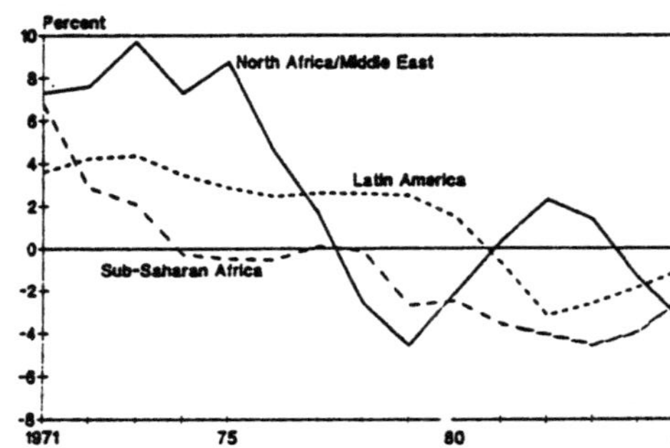

Figure 41
Annual change in real per capita income, South Asia, Southeast Asia, and Northeast Asia

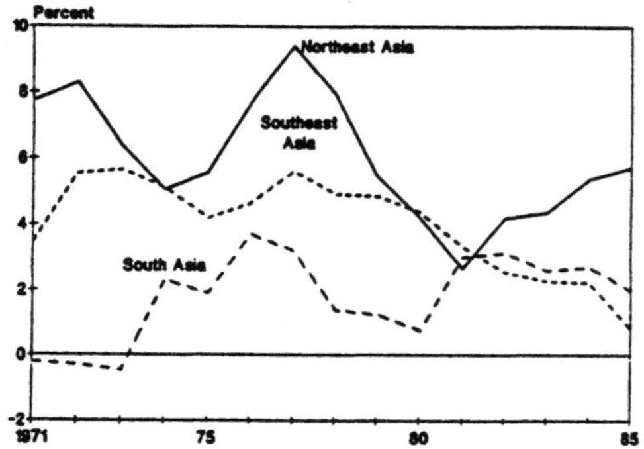

Figure 42
Trade in goods and nonfactor services, Sub-Saharan Africa

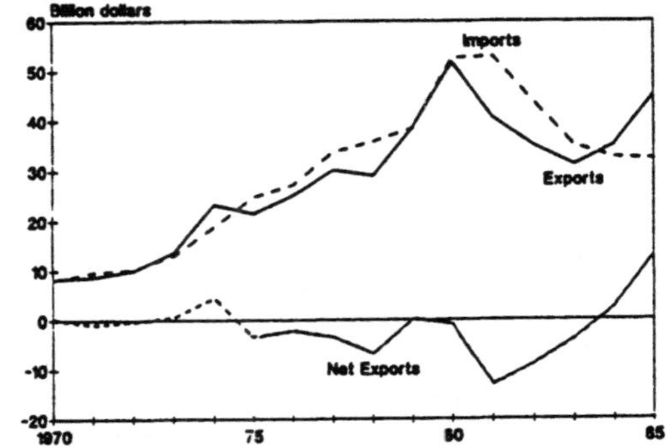

Figure 43
Trade in goods and nonfactor services, North Africa and the Middle East

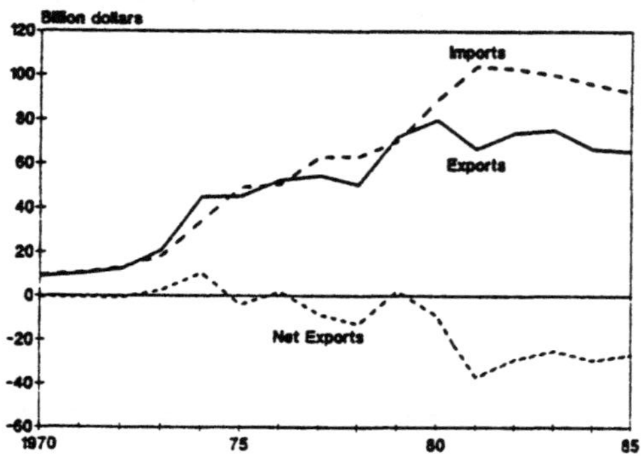

Figure 44
Gross domestic capital formation as a share of gross domestic product, all 79 countries, debt-affected countries, and Latin America

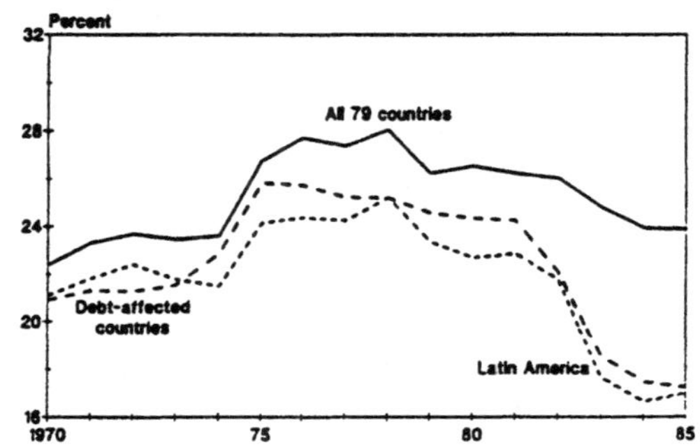

Figure 45

Gross domestic capital formation as a share of gross domestic products, South Asia, Southeast Asia, and Northeast Asia

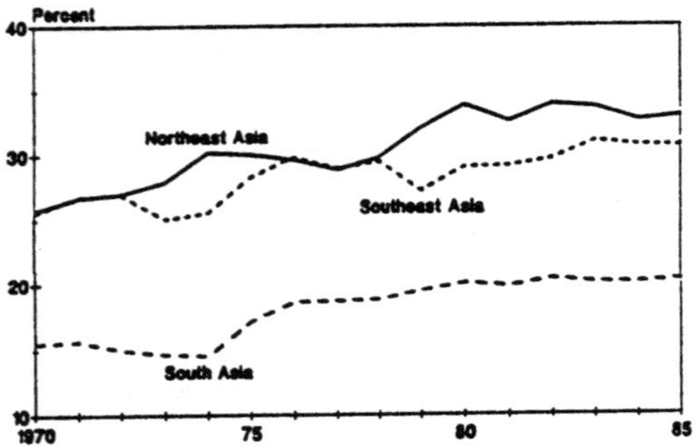

Figure 47

Agricultural trade, Southeast Asia

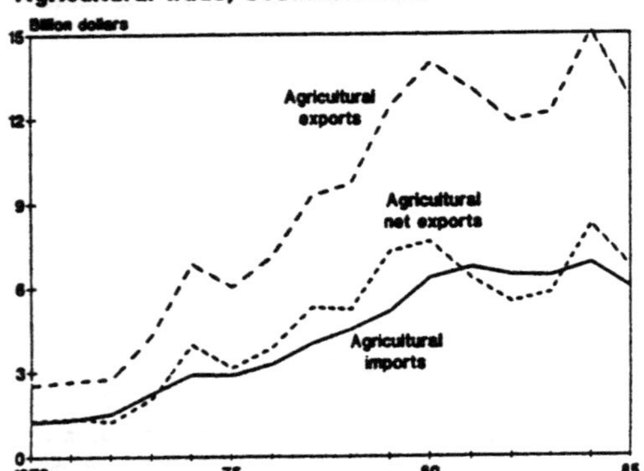

Figure 46

Agricultural trade, Sub-Saharan Africa

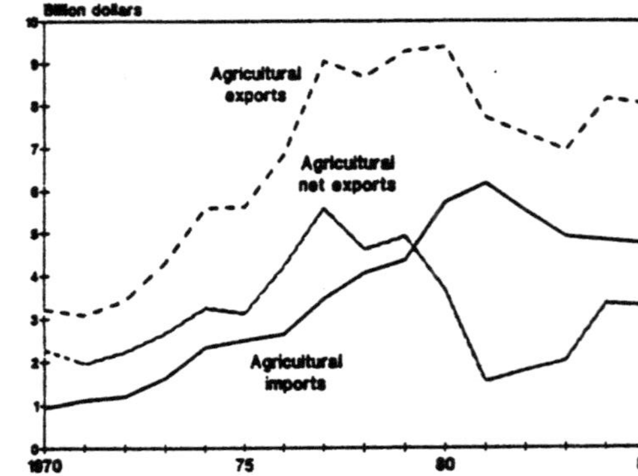

Figure 48

Agricultural trade, Northeast Asia

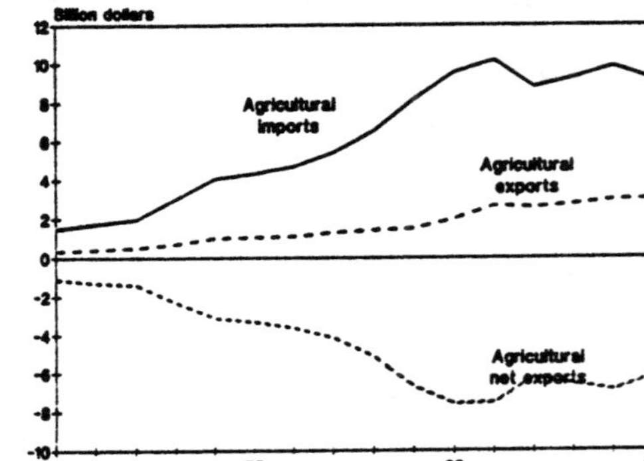

Figure 49
Agricultural trade, Latin America

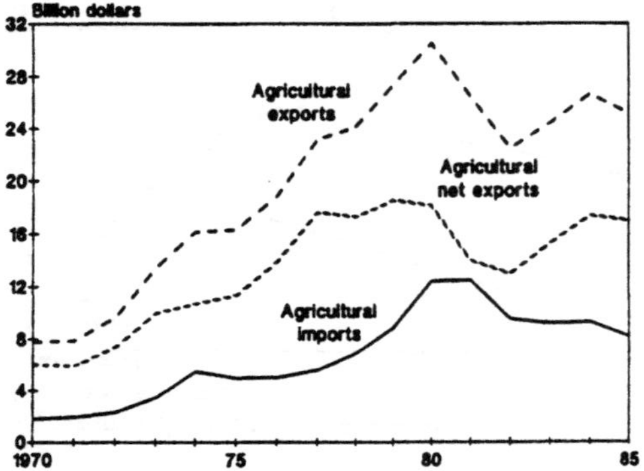

Figure 50
Agricultural exports as a share of all merchandise exports, all 79 countries

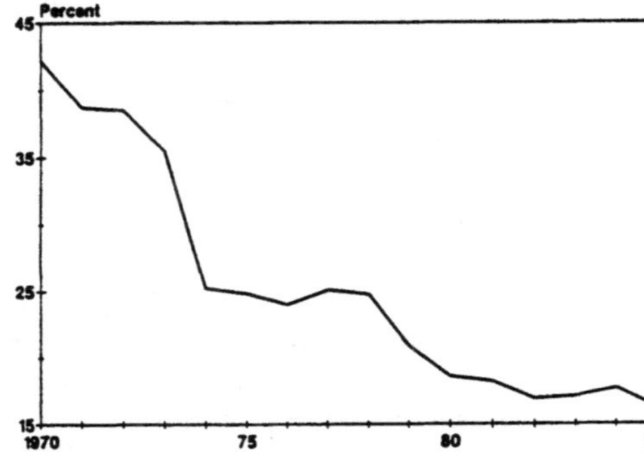

Figure 51
Agricultural exports as a share of all merchandise exports, Latin America and Sub-Saharan Africa

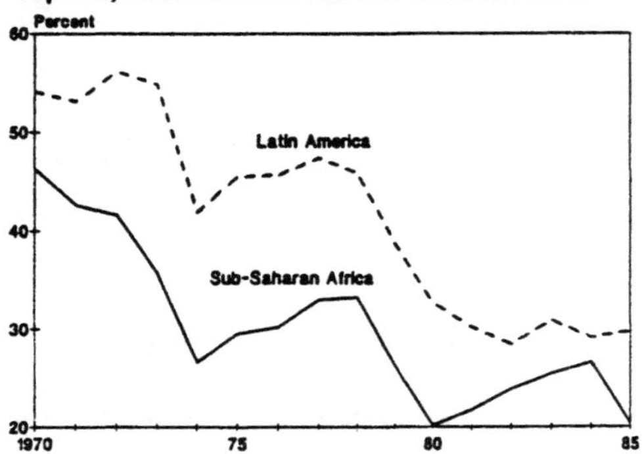

Figure 52
Gross domestic product growth under three scenarios, Latin America

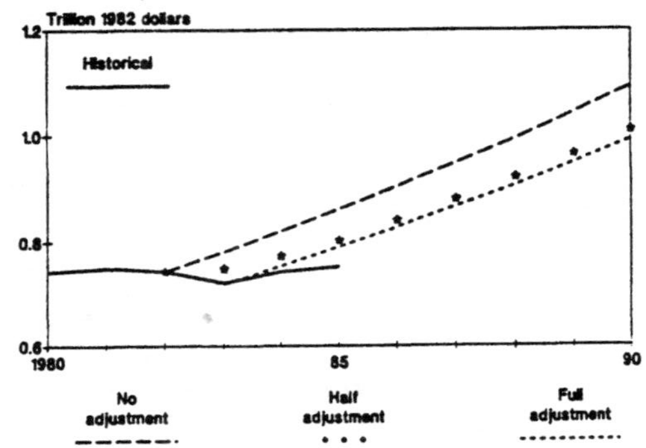

Figure 53
Import growth under three scenarios, Latin America

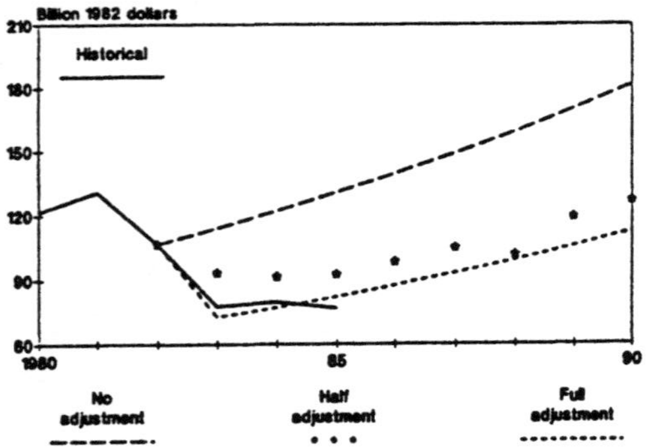

Figure 54
Gross domestic product growth under three scenarios, Sub-Saharan Africa

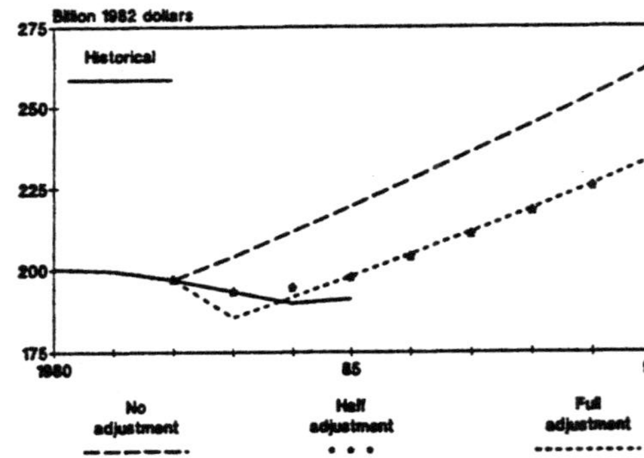

Figure 55
Import growth under three scenarios, Sub-Saharan Africa

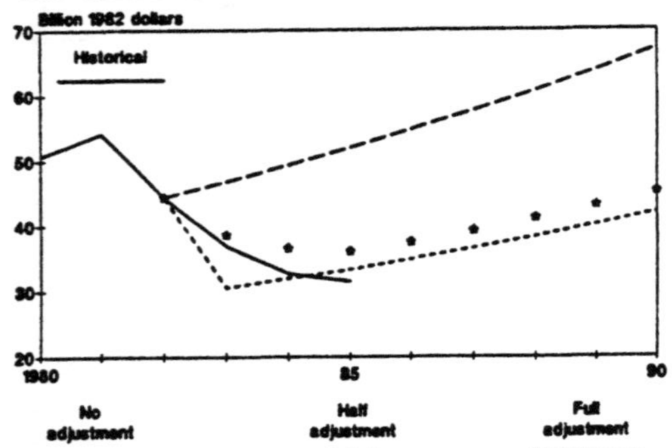

Figure 56
Gross domestic product growth under three scenarios, Northeast Asia

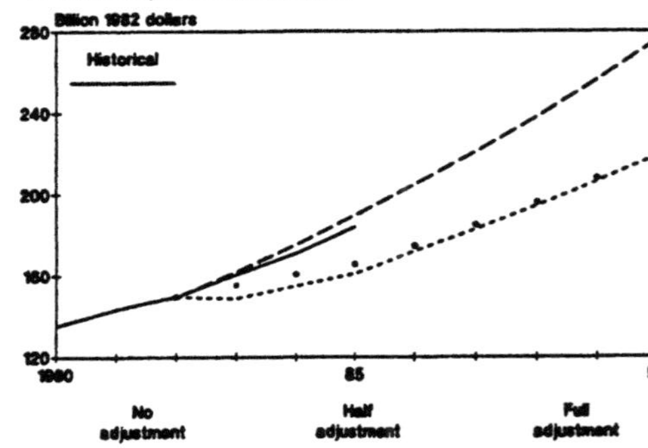

Figure 57
Gross domestic product growth under three scenarios, South Asia

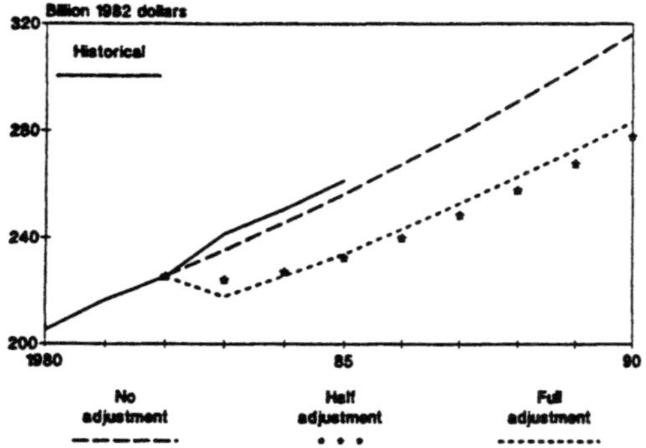

Figure 58
Import growth under three scenarios, Northeast Asia

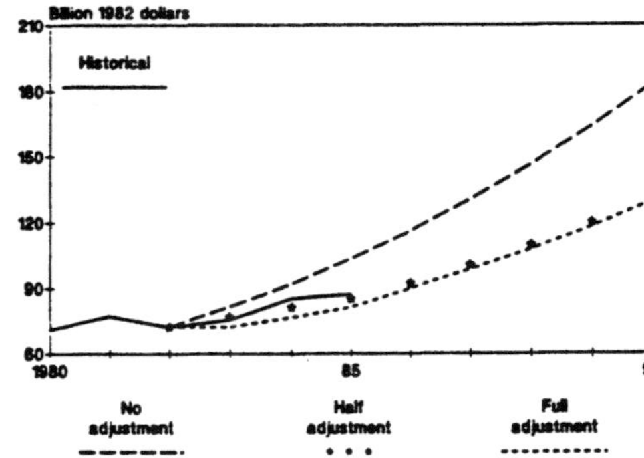

Figure 59
Import growth under three scenarios, South Asia

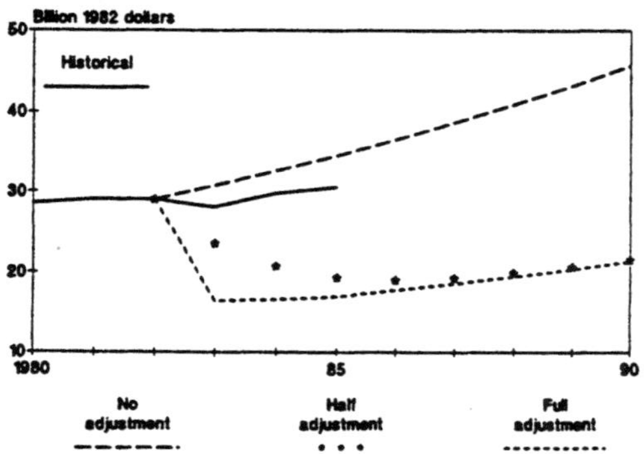

10
Modeling Exchange Rate and Macroeconomic Linkages to Agriculture: Lessons from a Structuralist Approach

Philip C. Abbott

INTRODUCTION

Theories of Exchange Rate Determination

Macroeconomic linkages to international trade, and in particular theories of exchange rate determination, have been in a state of considerable flux over the previous two decades. Early theory on the impact of exchange rates on trade emphasized relative price effects on the trade balance. This "elasticities approach" was based on movements in supply and demand of imports and exports as exchange rate changes altered the relative prices of tradeable commodities to non-tradeable commodities. McKinnon argued that in examining macroeconomic policy adjustments to exchange rate changes, this approach emphasized that policy makers consider the effect of changes in the trade balance on aggregate demand. Alexander's absorption approach explicitly recognized the importance of income effects arising from exchange rate changes. McKinnon indicates that this approach allows a separation of macroeconomic policy from exchange rate policy, which may be appropriate for an "insular" economy with limited linkages to the international economy, but it is inappropriate for today's more open, integrated world economy.

Modern exchange rate theory has added several additional considerations to earlier theory. They represent an integration of exchange rate theory with international macroeconomics. Purchasing

---

Philip C. Abbott, Professor, Department of Agricultural Economics, Purdue University, W. Lafayette, IN.

power parity, one of the first new theories, suggests that at least in the long run relative prices of commodities should equalize across countries. Hence, a relationship is developed between a change in exchange rates and the rate of inflation which accompanies it. The monetary approach takes purchasing power parity a step further. It recognizes that trade imbalance is countered by an offsetting international capital flow. These capital flows are recognized to be as important or more important than the trade balance in driving exchange rate movements. Parity conditions in this approach link interest rates across nations. The integration of international capital markets during the 1970s lent credence to this approach and to the assumptions underlying it. The portfolio approach modifies the interest rate parity conditions of the monetary approach by recognizing that assets of different nations are imperfect substitutes. Hence, interest rates adjusted for inflationary and exchange rate expectations need not be equalized across countries.

The success of these theories in explaining the historical record has been limited. Bilson and Martin report that most tests of the simplifying assumptions used in these models lead to their rejection. Neither purchasing power parity nor interest rate parity have fared well in those tests. Frankel reports that synthesis models incorporating the adjustment mechanisms of each model fare better than any of the simplified versions, but the predictive power of these synthesis models has also been questioned.

Structuralist Macroeconomics

Over this same period, another approach to international macroeconomics has developed. According to Krugman:

> Traditionally dominated by a historical and institutional approach, international monetary economics in the 1970s essentially became a branch of macroeconomics. This meant a drastic change in style....And the change in style meant a change in substance. What we know how to model formally are frictionless markets, where transactions are costless and agents make full use of the information available. The microeconomics of money, however, whether domestic or international, is fundamentally about frictions. (Paul Krugman, "The International Role of the Dollar: Theory and Prospect" in Exchange Rate Theory and Practice, J.F.O. Bilson and R.C. Martin, eds. Chicago: University of Chicago Press, 1984, pp. 261-2.)

This alternative approach has emphasized the institutions and adjustment processes of "institutional economists" and has been dubbed a structuralist approach. Taylor defines the structuralist approach as follows:

> An economy has structure if its institutions and the behavior of its members make some pattern of resource allocation and evolution substantially more likely than others. Economic analysis is structuralist when it takes these factors as the foundation stones for its theories.
>
> According to this definition, North Atlantic or neoclassical professionals do not practice structuralist economics. Their standard approach to theory is to postulate a set of interlocking maximization problems by a number of "agents" and ask about the characteristics of the solutions. Institutions are conspicuously lacking in this calculus, as is the recognition that men, women and children are political and social as well as economic animals. Non-economic, or even non-maximized forces affecting economic actions are ruled out of the discussion. Moreover, allowable economic actions are curiously circumscribed. For example, markets are almost always postulated to be price-clearing when it is patently obvious that many functioning markets are cleared by quantity adjustments or queues. (Lance Taylor, Structuralist Macroeconomics: Applicable Models for the Third World, New York: Basic Books, 1983, p. 3.)

One tenant of the structuralist approach is that real exchange rate changes, which change the relative prices of tradeables to non-tradeables, must alter the distribution of income. Political forces will resist this change, so that policies will be pursued to inflate away most real effects of a devaluation or other exchange rate change. In his review of the impacts of devaluation on 21 developing countries, Bautista found that inflation on average diminished by nearly 80% any nominal devaluations. Hence, in addition to the economic links between macroeconomic policy and trade of the more traditional exchange rate theories, this approach emphasizes a political dimension to that problem.

Corden also emphasizes the relationship between exchange rate changes and protectionism. He notes that as exchange rate appreciation increases competition from foreign imports, pressures for greater protection of domestic industry are likely to arise. McKinnon specifically cites the protection of agriculture as likely to lead to deviations in price adjustments from the consequences of exchange rate movements. Government intervention is recognized as a pervasive influence on agricultural trade (Dunmore and Longmire. A government's influence may profoundly affect the adjustment of relative prices in response to exchange rate movements. While inflation may erode much of the impact of a devaluation, agricultural prices, being subject to government control, may not change in accordance with overall inflation rates.

Mainstream macroeconomic theory and exchange rate modeling have also drawn from this structuralist school. Mark-up pricing rules which underlie the fixed price-flexible price characterization of a macroeconomy are decidedly structuralist in nature. This paradigm is particularly attractive to those modeling the linkages between the macroeconomy and agriculture. It is recognized that while agricultural and other commodity prices are determined in rapidly adjusting, competitive domestic markets, many manufactured goods and services prices are slow to adjust to changing market conditions, and are particularly unlikely to decline in the face of weak demand. As Taylor suggests, adjustment for these goods and services are more likely to be in quantity terms than in price.

AGRICULTURAL TRADE MODELS AND THE EXCHANGE RATE

Review

Recognition of a relationship between the exchange rate and performance in the agricultural sector came early in the development of the newer theories of exchange rate determination. Schuh first recognized the power of exchange rate changes to alter real agricultural prices in two settings--a developing country (Brazil) and in the United States. Following Schuh's work have been a number of attempts to refine the relationship between exchange rate changes and agricultural trade. The power of this mechanism as a driving force behind recent changes in international agricultural markets is increasingly recognized. In one of the more recent attempts to quantify the magnitude of these changes, Longmire and Morey attribute approximately $3 billion of the recent decline in U.S. wheat, corn and soybean exports to the appreciation of the dollar during the early 1980s.

Price Linkage-Policy Impact Paradox

Most approaches to exchange rate modeling in agricultural trade emphasize relative price adjustments (Collins, Meyers and Bredahl; Longmire and Morey). They operate in a partial equilibrium framework, not recognizing the essential distinction between traded and non-traded goods. Chambers and Just cite non-traded goods as an afterthought. While they claim their model can be adjusted to account for non-traded goods, that classification is not included in the specification reported in the paper. Income effects are only recently introduced to this discussion (Orden), and are not generally incorporated into trade models. While incorporation of inflationary adjustments is central to these approaches, they utilize variations on purchasing power parity and neutrality in inflation (with the possible exception of agriculture) to capture those real price changes. Price transmission elasticities are at times used to capture the effects of agricultural protectionism (or taxation) on this relative price adjustment.

These approaches raise a paradox which is especially evident in the policy conclusions drawn from these studies. Most agricultural trade models incorporate weak price responsiveness, especially on the part of importers, due to protectionism. It is sometimes argued that price policies, such as export subsidies or adjustments in loan rates and target prices, are likely to be ineffective due to this price inelasticity. Yet exchange rates, which in these models work through relative prices, are considered one of the main causes of the decline in U.S. agricultural exports. The model and policy conclusions of Longmire and Morey illustrate this paradox.

The answer to this paradox lies in the relationship between the exchange rate and macroeconomic outcomes, which in turn affects the levels of demand faced in international agricultural markets. While the exchange rate modelers recognize this deficiency and take it into account in their policy conclusions, it is not a part of standard agricultural trade models.

Assumptions typically invoked for the European Community also point to the importance of the relationship between the exchange rate and macroeconomic outcomes--the primacy of income effects. Domestic prices are generally assumed divorced from world market prices as a consequence of the Common Agricultural Policy. This cuts one of the main links through which exchange rate impacts on agricultural trade work. If exchange rate changes drive (or are symptoms of) changes in expenditure levels or foreign exchange allocations to trade or policy support (i.e., the budgetary allocation to the EC variable levy) this will in turn alter demand levels, even when relative domestic prices remain constant. This effect will vary by commodity, as it depends on income elasticities of demand and budget shares. These are generally very small for the grains--which are important to the U.S.--but may be larger for livestock products or other agricultural goods.

While much of the literature ignores certain aspects of the newer developments in theories of exchange rate determination, at least one application of each can be found. These refinements are not, however, adopted into the large trade models (GOL, IIASA) or into staff analysis of exchange rate impacts.

The relationship between macroeconomic policy and agricultural demand is taken into account in detailed models of the U.S. agricultural sector. International linkages of these models are often relatively weak, however. Freebairn, Rausser and de Gorter use the monetary approach to exchange rate determination in their model of these linkages and Chambers more recently used the portfolio approach. In both these models, however, the rest of the world is modeled as a single country. Differing degrees of price protection for importers and exporters are ignored. The world agricultural market is reduced to a single net export demand function. While this was a useful step in these models, it reduces their power to determine the magnitude of the impact of exchange rate changes (and of other macroeconomic policy with international repercussions).

Structuralist Alternatives to Agricultural Trade Modeling

The specifications of these models reflects a need by empirical modelers to adopt relatively simple modeling structures. Purchasing power parity has been an attractive alternative because of the modeling simplifications it allows. The interest rate parity condition of the monetary approach was adopted in several cases to achieve similar simplifications. The failure of these simplifications in empirical tests is disturbing to commodity modelers, who need to direct attention to the portions of models of direct concern, leaving formulation of exchange rate theories to experts in those areas. Structuralist assumptions may offer an alternative relevant to agricultural markets, where protection is important and where pricing rules and adjustment mechanisms may differ from those of other parts of the economy.

Agricultural economists are already incorporating "structuralist" concepts in agricultural trade models, and specifically in those models examining macroeconomic linkages to agricultural trade. Government intervention and protectionism are key aspects of agricultural trade models (Abbott (1979); Rausser, Lichtenberg and Lattimore). International institutions, such as GATT, and the primacy of domestic policy over international market behavior are hallmarks of the study of international agricultural trade. As Thompson argues, agricultural economists generally believe that inflation does not impact on agriculture in a manner similar to its impact elsewhere. Inflationary effects are not neutral. Hence, reliance on forms of purchasing power parity is inappropriate. Fixed price-flexible price models or the use of markup pricing for nonagricultural sectors are seen as more appropriate. Mark-up pricing rules have been used by Freebairn, Rausser and de Gorter, Shei and Thompson, and Chambers as important elements of the specification of macroeconomic linkages to agriculture.

If inflation is not neutral and prices do not bring about equilibrium, then disequilibrium frameworks may be better approaches. Models should concentrate on dynamic representations of adjustment processes in response to changing market conditions. Among the most important adjustment processes are agricultural policies which break the link between domestic and international prices and which serve to protect the agricultural sector.

The emergence of developing countries as importers in international agricultural markets suggests another concern for trade modelers (Abbott (1984b)). The mix of countries important to international capital markets and to currency adjustments do not correspond to those which matter to agricultural trade. The important international currencies are European or the yen, and these countries account for the bulk of international capital flows, as well. Over the 1970s and early 1980s, the developing countries and centrally planned economies were the dynamic elements in international agricultural markets of greatest importance to the U.S. For example, these two regions increased cereal imports by over 60 million metric tons, while imports of cereals by industrial market economies increased by only 0.6 million tons from 1974 to 1982 (FAO). The currencies of Eastern European countries and the USSR, as well as those of most LDC's, are inconvertible and not traded on currency markets. Many of these countries rely on the dollar as a vehicle for trade. These economies are not immune to the impacts of exchange rate and macroeconomic adjustment among the industrial market economies, however. Demand for exports, prices of imports, and international interest rates profoundly affect agricultural demand levels (Shane and Stallings).

What is most apparently missing from those models is a simple way to determine the "expenditure effects" of exchange rate and macroeconomic policy which determine the level of demand for agricultural commodities. By "expenditure effects" we mean to include both the effects of changing relative prices on income (and other avenues through which macroeconomic policy affects income) and the effect of capital flows, or foreign savings responsive to international interest rates, which drive a wedge between income and expenditure. While this is recognized by Orden, he offers no theory to be used in assessing the magnitude of this effect.

The central assertion of this paper is that we need to take greater guidance from the logic of the structuralist approach in future developments in agricultural trade modeling and in building macroeconomic linkages into those models. We believe that the most rapid progress in that effort will result from modifying existing modeling structures specifically designed for commodity markets, with those modifications guided by lessons from the experience of macroeconomists and international finance economists. The structuralist approach emphasizes that institutional factors specific to certain countries important to agricultural markets can be identified and incorporated in trade models. What must be found are behaviors and institutions which matter in economic adjustment--in this case specific to agricultural trade--which can serve as the basis for modifications to existing model structures.

Organization and Objectives

The task of the remainder of this paper will be to consider implications from the literatures on exchange rate modeling, structuralist macroeconomics, and agricultural trade modeling for modifying existing approaches to incorporating exchange rate adjustments and macroeconomic linkages into agricultural trade models. First, lessons from new developments in theories of exchange rate determination will be drawn. The boundaries to the agricultural trade modeling exercise will then be considered with respect to the needs for incorporating those lessons. In order to assess the relative importance of the issues raised, exchange rate linkages to agricultural trade will then be dissected into component parts, with the aim of limiting modifications of existing agricultural trade models to incorporation of those effects seen as being most important. Several avenues for future research on these macroeconomic linkages to agriculture will be identified in this process.

LESSONS FROM EXCHANGE RATE THEORY

Non-traded Goods and Inflation

The first lesson to be drawn from recent developments in exchange rate theory is the key role of non-traded goods and of induced inflation in determining the relative price changes which follow an exchange rate adjustment. General equilibrium trade theory shows that the primary effect of an exchange rate change on relative prices is to alter the ratio of traded to non-traded goods prices. Real prices do not generally adjust by the magnitude of the exchange rate change, however, as inflation raises the general price level by nearly the amount of the exchange rate adjustment. Hence, agricultural models focussing on the price effects of an exchange rate adjustment need to explicitly account for the pattern of real relative price changes actually going on.

A simple approach invoked in the past is to use purchasing power parity theory. In its simplest form, it suggests that an exchange rate movement will be matched by an equal change in the aggregate price level. Given its limited success in empirical tests, especially in the short to medium run, purchasing power parity is probably not the right assumption to invoke. Rather, empirical investigation of past responses of domestic inflation to real exchange rate adjustments on a country by country basis is a minimum approach. The alternative is to use some general equilibrium specification which incorporates both traded and non-traded goods.

Extensive protection of agriculture and isolation of domestic agricultural prices from border prices mean that agricultural prices are unlikely to follow adjustments in overall price levels. Institutions suggest that inflation's effect on agriculture is not neutral. While one would ordinarily include agricultural goods among traded goods in a general equilibrium specification, in many countries this would be the wrong approach. Especially where quantitative restrictions and non-tariff barriers are important, agricultural goods will behave like non-traded rather than traded goods. The structuralist approach would indicate that this dichotomy between traded and non-traded goods in standard trade models may not be appropriate either for certain policy regimes. For example, when the U.S. loan rate serves as an effective price floor, quantity rather than price adjusts to achieve an equilibrium. This fits neither the traded good nor the non-traded good specification of standard general equilibrium trade models.

Institutions like the U.S. loan rate policy, intended to achieve some stabilization of domestic agriculture, are probably more common in other countries than are open, free trade or fixed quotas--which could be equivalent to a non-traded good. That latter specification may be relevant where state trading is similar to a fixed quota regime, however.

Collins, Meyers and Bredahl suggest the use of price transmission elasticities to capture this adjustment of domestic agricultural prices to international price changes. They cite two models--stabilization of real domestic or nominal domestic prices--and show that the regime in place can make an enormous difference in the effect of an exchange rate. Nominal price stabilization suggests that the real change in agricultural prices is inversely proportional to the change in the price level, while real price stabilization suggests no change in real agricultural prices as the exchange rate changes. Neither regime is likely fully effective in practice, and the failure of purchasing power parity breaks the link between exchange rate adjustments and inflation. Collins' and Abbott's (1979) estimates of price transmission elasticities suggest that partial adjustment, and not these two extremes, is more likely in practice.

If exchange rate effects are to be isolated, two relationships need to be identified. One is the relationship between exchange rate adjustment and inflation, and the second is the relationship between domestic inflation rates and policy set agricultural prices. The international price transmission elasticity may only be a weak proxy for these two relationships. The power of the transmission elasticity approach will vary with the policy regime in place in a country. Our empirical understanding of price transmission effects is weak now. It may well be enhanced if we look more carefully at the relationship between domestic agricultural prices and inflation.

Macroeconomic Linkages

The above discussion of the importance of inflation suggests the second major lesson to be drawn from recent developments in theories of exchange rate determination. That is, the linkages between macroeconomic outcomes and exchange rate adjustments are important and need to be incorporated into the analysis of exchange rate impacts on agriculture. Three avenues of effect can be identified. Recent theory suggests that exchange rate movements and interest rate adjustments are closely related. Interest rate adjustments can significantly affect costs in agriculture, and yet these prices are left out of the partial equilibrium specifications. Expenditure levels also depend strongly on exchange rates, as foreign capital flows drive a wedge between income and expenditure through increased credit availability. These capital flows are now seen as being closely linked to exchange rate movements, and more importantly, monetary policy affecting domestic interest rates may be a driving force behind observed exchange rates. Income levels are also sensitive to exchange rates, either through changes in foreign demand induced by exchange rate changes or because policies, of which exchange rate adjustments are a part, may be contractionary or expansionary. Both of these last two mechanisms can significantly affect agricultural demand levels.

The linkages between interest rates and agriculture have been developed by Chambers; Rausser, de Gorter and Freebairn; and others. The primary relationship identified is between production costs, debt levels, and interest charges on loans to the agricultural sector. The effects of interest rates on storage costs and hence stockholding behavior is identified as another linkage. The international linkage has been handled in these studies through either the monetary approach to exchange rate determination or the portfolio approach. The latter, which assumes that assets of individual countries are imperfect substitutes, provides the basis for a reasonable specification of this linkage, if the trade relationships in Frankel's synthesis approach are not ignored.

Incorporating these relationships presents a difficult problem for the agricultural trade modeler who wishes to isolate the effects of exchange rate changes from other factors in a partial equilibrium model. The Chambers and Rausser, de Gorter and Freebairn approaches are general equilibrium and have relatively complicated specifications. They aggregate the rest of the world into one region. The alternative of building such complicated model structures for many regions is unrealistic, however.

At a minimum, agricultural sector specifications would need to add interest rates as prices in supply or net trade specifications to capture this linkage. Where stocks are important, interest rates should also be introduced as an argument of a stockholding behavioral function. Disentangling exchange rate influences on interest rates from influences arising from other causes would require specification of the relationship between interest rates and exchange rates for the major trading regions.

Institutional relationships may considerably simplify this task for many regions. For example, it is unlikely that most LDC's or centrally planned economies link domestic interest rates to international rates. Further, asset markets in these regions are seldom well developed and much investment is under state control. Given that currencies are inconvertible, the links specified in the portfolio approach are unlikely to hold for these regions. Separating the interest rate link between these regions and those where effective currency markets are likely to make this link important should help in assessing the magnitude of this link to agricultural trade. Not assuming that Canada, Japan and the European Community are a homogeneous region should also help the empirical performance of models examining trade. For those countries, institutional relationships may suggest simplified relationships between exchange rates and interest rates, as well.

Interest rate parity conditions offer such a simple relationship, which is more relevant to the developed countries than to LDC's or CPE's. The developed countries are where this linkage is strong, however. A careful look needs to be taken at the empirical tests of these conditions, to see what, if anything can be salvaged.

A second part of the new theories of exchange rate determination is the link between international capital flows and interest rates, which in turn depends on exchange rates. Trade imbalances are a fact of life, and are constantly changing for major traders. Assuming balanced trade will often miss an important effect on demand. Since the European Community and Japan are believed to isolate domestic prices from international prices, this linkage to demand levels may well dominate and can account for some of the observed changes in agricultural demand levels. That income elasticities and budget shares are low for some agricultural goods weakens this argument, however.

Developing this linkage through portfolio theory is probably inappropriate for the LDC's and centrally planned economies, however. Shane and Stalling's evidence suggest that foreign capital availability can significantly influence agricultural demand for those regions, but it also suggests that rationing by international financial institutions rather than demand for credit by those individual countries is the determining factor.

Structuralist representations of the behavior of international financial institutions is needed to set net foreign capital inflows to those regions. The LDC's and probably the centrally planned economies can then be modeled with those foreign capital inflows treated as being exogenously determined. Alternatively, they may depend on economic performance in those countries, which is the basis for the rationing rules.

Both expansion of expenditure over income, which results from foreign capital inflows, and income levels *per se* will affect agricultural demand levels. As Orden has recently argued, this linkage is generally overlooked in the literature relating agricultural trade to the exchange rate. An argument can be made that this mechanism dominates the relative price effects usually modeled, given the extent of price isolation which is typical of agricultural markets. This is especially true of countries in which policy sets real rather than nominal agricultural prices. In that case, border price changes would have no direct effect on the insulated commodity, but any induced changes on expenditure levels may still alter demand levels. Particularly for LDC's, modelers might focus efforts on the income linkages rather than price linkages.

Depending upon the macroeconomic policy regimes in place and the macroeconomic paradigm one holds to, this relationship between income and exchange rates can take several forms. Orden suggests a simple *ad hoc* specification which makes net foreign capital flows a function of relative prices. The rationale behind this is that changes in relative prices change the value for foreign exchange earnings relative to import costs and hence changes income. While careful accounting of border price adjustments and their implications for net trade balances is needed, two additional relationships may be identified, however. One is that there may be a multiplier effect through a Keynesian impact on demand from higher or lower net trade. This is essentially Alexander's absorption approach to exchange rate adjustment. The second is that internal macroeconomic adjustments to the inflation and interest rate changes induced by an exchange rate movement may alter income. These latter influences may be better represented by a direct relationship between exchange rates (and/or international interest rates) and income.

Orden's approach would also need to be augmented by any relationship between exchange rates and rationing of foreign capital flows. This suggests the use of a relationship between expenditure rather than income and exchange rates. Again, a structural representation of these links through specification of a simple macroeconomic model based on existing institutions and policy may lead to a better understanding of this relationship than will estimation of the reduced form relationship between exchange rates, interest rates and expenditure levels.

BOUNDARIES ON THE MODELING EXERCISE

The lessons discussed above have led to suggestions on limitations to the modeling of exchange rate linkages to agricultural trade. While the boundaries to such an exercise will of necessity depend on research issues more specifically defined, some further lessons on modeling approaches may be drawn from those considerations. The two main concerns addressed here are whether general equilibrium modeling is necessary (or can modified partial equilibrium suffice), and what extent of country coverage is necessary for the rest of the world (ROW). Empirical problems in estimating the necessary relationships are relevant to this latter discussion. It is presumed that the spatial equilibrium model of Collins, Meyers and Bredahl serves as the starting point for this discussion.

General or Partial Equilibrium

Whether general equilibrium modeling is necessary or not depends upon the magnitude of feedback linkages between agriculture and the rest of the economy. While the literature is clear on the importance of macroeconomic factors affecting agricultural outcomes, evidence on impacts in the other direction is less clear. If agricultural prices and demand for expenditure and credit do not significantly impact the overall price level, interest rates, or income and expenditure, then treating these variables as exogenous to the trade modeling exercise will suffice. Given the need to simplify the modeling exercise to a manageable level, and given the potential for incompatibility between standard commodity models and macroeconomic models, this one way linkage may well suffice. Forward linkages from the macroeconomy to agriculture can be handled by letting macroeconomists deal with relationships between exchange rates, interest rates, inflation rates and expenditure. These are then exogenously entered as shifters of supply, demand and trade in the agricultural models. The importance of treating the backward linkages from agriculture is an unsettled question, however. It has been argued, for example, that food prices play a disproportionately large role in the formation of inflationary expectations, although evidence on that is inconclusive (Bilson and Martin).

Our position is that unless the emphasis of the research is on separating the effects of exchange rates from other macroeconomic impacts, ignoring the backward linkages and sticking to partial equilibrium frameworks that add variables called for by the adjustment mechanisms identified above may suffice. Since the empirical evidence suggests that models of exchange rate determination are incomplete, and given the highly interrelated nature of the macroeconomic variables in question, attempting that separation is likely to be a fruitless exercise at present. Whether the exchange rate is a primary cause of the decline in agricultural markets or is a symptom of underlying macroeconomic causes is not as important to identify as is the ability to assess the consequences of exchange rate and accompanying macroeconomic changes on agricultural markets.

Detail in ROW

In considering how large to make these models, the general equilibrium-partial equilibrium debate emerges again. One can identify two extreme approaches to this question. On the one hand, the export market faced by the U.S. is often modeled as an aggregate rest of the world. The alternative is to model the behaviors of many countries or regions. Both approaches face limitations from the weak empirical basis for the world commodity market models (Thompson).

Estimates of an elasticity of net agricultural export demand vary over a wide range (Gardiner and Dixit) and are extremely difficult to estimate (Orcutt). Given the number of actors in these markets with differing behaviors affected by several variables, specification error in the direct estimation of these functions is inevitable. Imperfect competition exercised by government agencies in these markets may mean that a stable net export demand function may not exist. Synthetic procedures based on supply and demand in individual countries important to a particular market have been used as an alternative estimation procedure. This approach is followed by Freebairn, Rausser and de Gorter. Problems with this approach stem from our limited knowledge of supply and demand elasticities--often the same elasticities must be assumed for a wide range of countries--and more importantly our lack of knowledge of price transmission parameters. These are often based simply on institutional arguments. While Collins, Meyers and Bredahl observe that nominal and real price protection lead to very different price and exchange rate adjustments, we seldom know which regime applies in a particular country.

Attempts at directly estimating the relationship between the exchange rate and agricultural trade face these same econometric problems. As Belongia observes, conclusions concerning the impact of exchange rates on agricultural trade can be reversed by simply choosing a different index to serve as the basis for setting the magnitude of the exchange rate used in the evaluation. This should hardly be surprising, however. We know that when the U.S. exchange rate changes relative to the S.D.R., exchange rates of other countries will also adjust, but not uniformly. That there are several exchange rate indices which differ depending on weights assigned to different currencies suggests that these currency movements are seldom uniform across the currencies which matter to agricultural trade. What we know about price transmission suggests that exporters adjust domestically more than importers to world price changes--with the exception of the European Community. Hence, an adjustment in the U.S. exchange rate relative to the mark may have little effect on European agricultural prices, and the effect on Canadian prices depends on whether the Canadian dollar followed the mark or not.

The Orcutt and Gardiner and Dixit arguments suggest that we are unlikely to be very successful in direct estimation of the necessary behavioral relationships for an aggregate ROW, given the complexity of the international market and the potential for specification error bias. There is little reason to suspect that we should be more successful in estimating the reduced form exchange rate impact as proposed by Chambers and Just. Belongia's evidence supports this fear that direct estimation of exchange rates impacts is unlikely to lead to robust, stable empirical results.

On the other hand, building agriculture cum macro models on the scale of those built for the U.S. by Chambers and by Rausser, de Gorter and Freebairn for all countries important to agricultural trade is unrealistic. Those models which do choose to incorporate several regions are often out of date from changes in the markets which are now most important to U.S. agricultural exports. More emphasis needs to be placed on the LDC's and centrally planned regions than is now the case. In current models, LDC's are often lumped into a single residual category with no attempt made to look at exchange rate effects in those regions. The theory used in those models is also generally better adapted to the developed countries than to centrally planned or developing countries. As the income and foreign capital effects on these countries can be substantial, their inclusion will enhance our understanding of the quantitative significance of these linkages. Estimation of the needed empirical relationships is less readily available, however, and data quality in the LDC's is weak.

DISSECTING EXCHANGE RATE IMPACTS

The philosophy behind any structuralist approach to modeling exchange rate impacts on and macroeconomic linkages to agriculture involves identification of the important relationships and using institutional arrangements to specify and estimate those relationships. The discussion of modifications to existing approaches by looking at the significant relationships arising from newer theories of exchange rate determination and from international macroeconomics followed that approach.

An alternative way of approaching this question is to dissect the linkages between agricultural trade and exchange rate movements into component parts. This section attempts such a dissection as a means of identifying the critical linkages. Three separate linkages are identified:

1. Exporter competitiveness adjustments
2. Price induced adjustments in importer demand
3. Income and expenditure effects on demand

These linkages may be identified with relevant parts of the above discussion.

Exporter Competitiveness

Exchange rate changes among the major competing agricultural exporters can significantly influence the competitiveness of one country relative to another. In grains markets, for example, those stocks which are in excess of pipeline needs are generally held by a few major exporters. Further, the assumptions generally made concerning price transmission set exporters as relatively open to international agricultural markets, whereas importers exert considerably greater insulation. Nevertheless, importers may choose to import the fixed quantity in question (assuming other exchange rate linkages are also inoperative) from the cheapest source. As exchange rates adjust, the cost of an exporter with a devalued currency will decrease relative to that of a country whose currency is appreciating. While market adjustments may bring those prices back together, in so doing it will allow the country whose exchange rate depreciated to capture greater market share and increased exports.

Adjustments may come in stocks positions, domestic demand or in real farm income levels. Bilateral exchange rates between the significant exporters is key to this "competitiveness" relationship. Further, as domestic demand adjustments are likely to be small, given low elasticities in most developed countries, the key exchange rates are those between exporters taking stocks positions. In wheat, the U.S.-Canadian exchange rate may be the key relationship. European rates are less important because agricultural policies lead to export levels unresponsive to prices. Similarly, Argentine and Australian exchange rates may hold less significance for this "competitiveness effect" because of price insulation due to policy in the former and the lack of stocks through which rapid quantity adjustments may be made in both. The effect on income levels of farmers from these price adjustments, where they are reflected in the longer run in domestic prices, may induce supply response, however.

The above arguments were based on the assumption of a competitive international market for agricultural commodities. Recent evidence provided by Figueroa suggests that this may not be the case. His estimates of trade flow patterns based on an Armington model structure indicate that rigidities in marketing relationships, bilateral agreements, reliability of exporters, and fundamental differences in quality characteristics of agricultural commodities across exporters lead to persistence in existing patterns. Kolstad and Burris cite imperfect competition as another factor leading to more diverse but persistent market shares for exporters. While this suggests that trade flow patterns are slower to adjust to relative price changes at exporter borders, those changes can persist, leading to competitive advantage for particular exporters. Especially in the medium to longer run, this suggests that exchange rate adjustments will not be eliminated by competitive market response. In the competitive model and without stocks, adjustments are likely to be in real farm incomes simply through price changes relative to inflation. The Armington structure and imperfect competition models are more likely to lead to real price and quantity shifts.

Prices and Imports

Price effects on importers are also worth considering separately from exporter responses. Price transmission models are used to gauge the impact on the domestic sector of international price changes. It is not uncommon to assume most importers, especially LDC's and centrally planned economies, insulate domestic prices from international prices. This behavior is in sharp contrast to the usual assumption for exporters. In many of those countries, price controls are not limited to agricultural goods, but may be widespread throughout the economy. If that is the case, little effect through price on demand is likely.

The key to this relationship is the extent of inflation in response to exchange rate movements and the nature of the agricultural policy regime--are prices set nominally or in real terms. The discussion earlier of prices and inflation is probably more appropriate to these importers than to exporters. For exporters, the significant effect is the competitiveness effect, while for importers the central question is how do relative price shifts alter domestic demand and supply and how are those price shifts dependent on exchange rate adjustments.

Expenditure Effects

Income effects and expenditure adjustments may be more important than relative price adjustments, however. Even if a country insulates domestic prices from international prices, it may well not avoid the effects of macroeconomic linkages on demand. These will affect the developed and developing countries alike, but through different mechanisms. Since the income elasticities of demand and agricultural budget shares for most developed countries are believed to be very low, even if exchange rate adjustments have significant effects on expenditure, their effect on demand may be small and, if so, could be ignored. In many LDC's however, income effects are seen to be crucial. They are especially important as a country passes through a transition phase from a grain based diet to one incorporating more animal products (Mellor and Johnson). Both income elasticities of demand and budget shares on agricultural goods are likely to be larger for these countries. Both these factors lead to greater impacts from exchange rate shifts and accompanying macroeconomic adjustments.

International financial adjustments must also be better understood if the "expenditure effect" from exchange rate adjustment is to be captured. The determination of the magnitude of foreign capital to LDC's and its distribution among those countries needs to be better understood. This aspect has been virtually ignored, even in those papers which address the implications for agricultural trade of the debt crisis (Shane and Stallings). The contraction in foreign capital availability in those models is exogenous, while its impact is seen as being very significant.

Foreign exchange may also play a greater role in determining agricultural trade levels than is suggested by expenditure changes and consequent adjustments in agricultural demand. State involvement in the control of foreign exchange allocations is important in both LDC's and centrally planned countries. Quantitative restrictions, especially when foreign exchange is tight, are not uncommon. Agricultural imports compete with priority imports of capital and intermediate goods. Models of foreign exchange allocation, such as those of Scobie or Hjort and Abbott, offer a starting point for modeling this institution.

The "structuralist" implications of the above discussion are that two issues should be the focus of efforts to modify existing modeling structures. The price effect on importer demand is probably the best handled issue currently. More attention needs to be paid to "competitiveness" of exporters and particularly the market structure within which this is determined. Also, how foreign capital flows are determined and allocated among imports by LDC's is a crucial consequence of exchange rate adjustment and its macroeconomic consequences.

CONCLUSIONS

The purpose of this paper is to argue that our understanding of international agricultural markets, coupled with an understanding of the mechanisms through which the exchange rate impacts an economy, may be used to identify modifications to existing approaches to modeling the linkages between agricultural trade and the exchange rate. The logic of the "structuralist" approach to macroeconomics has been helpful in this exercise. Two approaches to identifying the relevant structure and institutions of agricultural markets and their implications for modeling have been followed. First, lessons from recent developments in theories of exchange rate determination were related to the literature on exchange rate impacts on agricultural trade. The exchange rate impacts were then dissected into component parts.

Several conclusions emerge concerning appropriate approaches to modeling this linkage to agricultural trade. The paradox that exchange rate impacts are typically modeled through price linkages, whereas insulation of agriculture may limit this price linkage was considered. The inflationary impact of a devaluation and the agricultural policy regime in place emerge as key concerns.

That the impacts on income and expenditure may dominate due to agricultural price insulation was also considered. The importance of expenditure effects points to the need to understand the relevance of income effects on demand, foreign exchange allocation by LDC's, and rationing of foreign capital flows by international financial institutions. Careful accounting of the income effects are important, and so are the macroeconomic consequences of changes in a country's budget constraint resulting from international price shifts. Government budgetary policy is related to this link and to foreign capital flows of both developed and developing countries.

The competitiveness effect of exchange rates altering the source of cheapest imports was also suggested as a concern meriting further attention. Two arguments lead to this concern. The diversity and persistence of market shares among exporters suggests that a competitive spatial equilibrium may be an inadequate representation of the operation of international agricultural markets. Differentiation of products by exporter, as assumed in the Armington model structure, argues for the importance of marketing rigidities, bilateral trade agreements, and the non-homogeneity of product characteristics. Imperfect competition among exporters may also lead to this observed pattern of trade. The competitiveness mechanism operates even if protection of agriculture reduces price responsiveness of importers. It depends largely on the price responsiveness of exporters, which is strongly affected by stockholding behavior.

The above issues may be addressed through modification of existing agricultural trade model structures. A full general equilibrium treatment linking macroeconomic and commodity models is needed only if exchange rate and other macroeconomic impacts must be disentangled. Given the controversies which persist in international macroeconomics, that disentanglement promises to be a difficult task.

REFERENCES

Abbott, P. C. Foreign Exchange Constraints to Trade and Development. FAER No. 209. Economic Research Service, U. S. Department of Agriculture. Washington, D.C., November, 1984a.

_______. "Government Policies, Foreign Debt and the Prospects for Agricultural Trade and Economic Development of a Low Income African Country--the Case of Guinea Bissau." Presented at the AAEA annual meeting, Ithaca, NY, August 1984b.

_______. "Modeling International Grain Trade with Government Controlled Markets," American Journal of Agricultural Economics. 61(1979): 22-31.

Alexander, S. S. "Devaluation versus Import Restrictions as an Instrument for Improving Foreign Trade Balance," International Monetary Fund Staff Papers. 1(1951): 379-396.

Bautista, R. M. "Exchange Rate Adjustment Under Generalized Currency Floating: Comparative Analysis Among Developing Countries." World Bank Staff Working Paper No. 436. Washington, D.C. October, 1980.

Belongia, M. T. "Estimating Exchange Rate Effects on Exports: A Cautionary Note," Federal Reserve Bank of St. Louis Review. 68(1) 1986. pp. 5-16.

Bilson, J. F. O. and R. C. Martin. (eds.) Exchange Rate Theory and Practice. Chicago: University of Chicago Press, 1984.

Chambers, R. G. "Agricultural and Financial Market Interdependence," American Journal of Agricultural Economics. 66(1984): 12-24.

_______, and R. Just. "A Critique of Exchange Rate Treatment in Agricultural Trade Models." American Journal of Agricultural Economics. 61(1979): 249-57.

_______, and R. Just. "Effects of Exchange Rate Changes on U.S. Agriculture: A Dynamic Analysis." American Journal of Agricultural Economics. 63(1981): 32-46.

Collins, H. C. "Price and Exchange Rate Transmission," Agricultural Economics Research. 32(1980): 50-55.

Collins, K. J., W. H. Meyers, and M. Bredahl. "Exchange Rates and Agricultural Prices," American Journal of Agricultural Economics. 62(1980): 656-65.

Corden, W. M. "Protection, the Exchange Rate and Macroeconomic Policy," Finance and Development. 22(1985): 17-19.

Dornbusch, R. Open Economy Macroeconomics. New York: Basic Books, 1980.

Dunmore, J. and J. Longmire. "Sources of Recent Changes in U.S. Agricultural Exports," ERS Staff Report No. AGES831219. Economic Research Service, U.S. Department of Agriculture. Washington, D.C., January 1984.

_______. Food Policies in Developing Countries, FAER No. 194. Economic Research Service, U.S. Department of Agriculture. Washington, D.C, December, 1983.

Figueroa, E. "The Impacts of Movements in U.S. Exchange Rates on Commodity Trade Patterns and Composition." Ph.D. thesis, University of California-Davis. December 1986.

Frankel, J. "Tests of Monetary and Portfolio Balance Models of Exchange Rate Determination," Exchange Rate Theory and Practice. J. Bilson and R. Marston, (eds.). Chicago: University of Chicago Press, 1984.

Freebairn, J. W., G. C. Rausser and H. de Gorter. "Monetary Policy and U.S. Agriculture," International Agricultural Trade: Advanced Readings in Price Formation, Market Structure and Price Instability. G. Storey, A. Schmitz and A. Sarris, (eds.), Boulder, CO: Westview Press, 1984.

Gardiner, W. H. and P. M. Dixit. "Price Elasticity of Export Demand: Concepts and Estimates," ERS Staff Report No. AGES 860408. Economic Research Service, U.S. Department of Agriculture. Washington, D.C., May, 1986.

Hjort, K. and P. C. Abbott, "Soviet Plans and Economic Performance: Implications for Grain Trade," presented at the IATRC annual meeting. Vancouver, B.C. December, 1985.

Kaldor, N. "Devaluation and Adjustment in Developing Countries," Finance and Development. 20(1983): 35-37.

Kolstad, C. D. and A. E. Burris. "Imperfectly Competitive Equilibria in International Commodity Markets," American Journal of Agricultural Economics. 68(1986): 27-36.

Krugman, P. "The International Role of the Dollar: Theory and Prospect," Exchange Rate Theory and Practice. J. F. O. Bilson and R. C. Martin (eds). Chicago: University of Chicago Press, 1984. pp. 261-2.

Longmire, J. and A. Morey, Strong Dollar Dampens Demand for U.S. Farm Exports, FAER No. 193. Economic Research Service, U.S. Department of Agriculture. Washington, D.C. December, 1983.

McKinnon, R. I. "The Exchange Rate and Macroeconomic Policy: Changing Postwar Perceptions," Journal of Economics Literature. XIX, (1981): 531-57.

Mellor, J. W. and B. F. Johnson. "The World Food Equation: Interrelations Among Development Employment and Food Consumption." Journal of Economic Literature. XXII(1984): 531-74.

Orcutt, G. H. "Measurement of Price Elasticities in International Trade," Review of Economics and Statistics. 32(1950): 117-32.

Orden, D. "A Critique of Exchange Rate Treatment in Agricultural Trade Models: Comment." American Journal of Agricultural Economics. 68(1986): 990-93.

Rausser, G. C., E. Lichtenberg, and R. Lattimore, "Developments in Theory and Empirical Applications of Endogenous Government Behavior," G. C. Rausser, ed. New Directions in Modeling and Forecasting in U.S. Agriculture, Amsterdam: North Holland, 1983.

Schuh, G. E. "The Exchange Rate and U.S. Agriculture," American Journal of Agricultural Economics. 56(1974): 1-13.

Scobie, G. Government Policy and Food Imports: The Case of Wheat in Egypt, IFPRI Report No. 29. Washington, D.C. December, 1981.

Shane, M. D. and D. Stallings, "Trade and Growth of Developing Countries Under Financial Constraint," ERS Staff Report No. AGES840519. Economic Research Service, U.S. Department of Agriculture. Washington, D.C., June, 1984.

Shei, S. Y. and R. L. Thompson, "Inflation and Agriculture: A Monetarist-Structuralist Synthesis," Dept. Agr. Econ. Working Paper. Purdue Univ., W. Lafayette, IN, 1981.

Taylor, L. Macro Models for Developing Countries, New York: McGraw-Hill, 1979.

_______. Structuralist Macroeconomics: Applicable Models for the Third World, New York: Basic Books, 1983.

Thompson, R. L. A Survey of U.S. Developments in International Agricultural Trade Models, Bibliographies and Literature of Agriculture No. 21. Economic Research Service, U.S. Department of Agriculture. Washington, D.C., September, 1981a.

_______. "On the Power of Macroeconomic Linkages to Explain Events in U.S. Agriculture: Discussion." American Journal of Agricultural Economics. 63(1981b): 888-9.

For Product Safety Concerns and Information please contact our EU representative GPSR@taylorandfrancis.com Taylor & Francis Verlag GmbH, Kaufingerstraße 24, 80331 München, Germany

Batch number: 10725927

Printed by Printforce, the Netherlands